STRUCTURE–SOLUBILITY RELATIONSHIPS IN POLYMERS

Academic Press Rapid Manuscript Reproduction

Proceedings of the Symposium on
Structure-Solubility Relationships in Polymers
held by the American Chemical Society in
San Francisco, California, on August 30-September 3, 1976.

STRUCTURE-SOLUBILITY RELATIONSHIPS IN POLYMERS

EDITED BY

Frank W. Harris

Wright State University
Dayton, Ohio

Raymond B. Seymour

University of Southern Mississippi
Hattiesburg, Mississippi

ACADEMIC PRESS New York San Francisco London 1977
A Subsidiary of Harcourt Brace Jovanovich, Publishers

ACADEMIC PRESS, INC.
111 Fifth Avenue, New York, New York 10003

United Kingdom Edition published by
ACADEMIC PRESS, INC. (LONDON) LTD.
24/28 Oval Road, London NW1

Library of Congress Cataloging in Publication Data

Main entry under title:

Structure-solubility relationships in polymers.

Includes index.
Papers presented at a symposium held by the American Chemical Society at its 172d meeting in San Francisco, Aug. 30–Sept. 3, 1976, and sponsored by the Divisions of Polymer Chemistry and Organic Coatings and Plastics Chemistry.
1. Polymers and polymerization–Congresses. 2. Solubility–Congresses. I. Harris, Frank Wayne, Date II. Seymour, Raymond Benedict, Date III. American Chemical Society. Division of Polymer Chemistry. IV. American Chemical Society. Division of Organic Coatings and Plastics Chemistry.
QD381.8.S76 547′.84 76-58386
ISBN 0–12–327450–8

PRINTED IN THE UNITED STATES OF AMERICA

DEDICATION

It is a distinct and monumental occasion when a former president of the American Chemical Society serves as the keynote speaker for a symposium which is based on some of his original investigations. It certainly is in accord with this Centennial/Diamond Jubilee meeting when this 94 year old patriarch presents the principal talk at the president's dinner and the opening address of the symposium on the structure-solubility relationships in polymers.

His utilization of Occam's razor to reduce the complexity of solubility theories and his "From Then to Now" principal address are most appropriate for a Centennial meeting of a chemical society in which he has played a major role throughout his professional lifetime. We are confident that the authors of the following papers join with the chairmen in dedicating this book to Dr. Joel H. Hildebrand who is responsible for much of the progress that has been made in relating the solubility of polymers to their molecular structure.

CONTENTS

LIST OF CONTRIBUTORS

RONALD H. BANEY (225), DOW Corning Corporation, Corporate Research, Mail #66, Midland, Michigan 48640

G. C. BERRY (71), Department of Chemistry, Carnegie-Mellon University, Pittsburgh, Pennsylvania 15213

R. F. BLANKS (111), Department of Chemical Engineering, Michigan State University, East Lansing, Michigan 48824

D. C. BONNER (233), Department of Chemical Engineering, Texas A and M University, College Station, Texas 77843

J. A. CAROTHERS (11), IBM Research Laboratory, Monterey and Cottle Roads, San Jose, California 95193

CHARLES E. CARRAHER, JR. (215), Department of Chemistry, Wright State University, Dayton, Ohio 45431

K. C. CHEN (233), Department of Chemical Engineering, Texas Tech University, Lubbock, Texas 79409

J. DERUDDER (59), Materials and Metallurgical Engineering and Macromolecular Research Center, University of Michigan, Ann Arbor, Michigan 48109

MALCOLM DOLE (123), Department of Chemistry, Baylor University, Waco, Texas 76703

PAUL L. DUBIN (135), Dynapol, 1454 Page Mill Road, Palo Alto, California 94304

F. E. FILISKO (59), Materials and Metallurgical Engineering and Macromolecular Research Center, University of Michigan, Ann Arbor, Michigan 48109

L. W. FISHER (149), Research Laboratories, Eastman Kodak Company, Rochester, New York 14650

J. GOLBA (59), Materials and Metallurgical Engineering and Macromolecular Research Center, University of Michigan 48109

FRANK W. HARRIS (183), Department of Chemistry, Wright State University, Dayton, Ohio 45431

JOEL H. HILDEBRAND (1), Department of Chemistry, University of California, Berkeley, California 94720

EARL J. JOHNSON (241), Baylor College of Medicine, Houston, Texas 77002

H. T. KAU (59), Materials and Metallurgical Engineering and Macromolecular Research Center, University of Michigan, Ann Arbor, Michigan 48109

MARK E. LAND (123), Department of Chemistry, Baylor University, Waco, Texas 76703

LYNN H. LANIER (183), Department of Chemistry, Wright State University, Dayton, Ohio 45431

O. C. C. LIN (165), E. I. duPont de Nemours and Company, Marshall Laboratory, P.O. Box 3886, 3500 Grays Ferry Ave., Philadelphia, Pa. 19146

JAMES W. MENTELE (225), DOW Corning Corporation, Corporate Research, Mail #66, Midland, Michigan 48640

KENNETH D. NISBET (33), Technical Coatings, Inc., P.O. Box 5427, Lubbock, Texas 79417

A. C. OUANO (11), IBM Research Laboratory, Monterey and Cottle Roads, San Jose, California 95193

DONALD R. OWEN (245), Department of Materials Science, University of Utah, Salt Lake City, Utah 84112

JOSÉ PÉREZ B. (89), Departimento de Química, Universidad Católica Andrés Bello, Caracas, Venezuela

H. ROSENBERG (233), Air Force Materials Laboratory, Polymer Branch, Wright-Patterson AFB, Ohio 45433

ARTURO RUBIO (89), Departimento de Química, Universidad Católica Andrés Bello, Caracas, Venezuela

RAYMOND B. SEYMOUR (99, 241, 259), Department of Polymer Science, University of Southern Mississippi, Hattiesburg, Mississippi 39401

B. N. SHAH (111), Department of Chemical Engineering, Michigan State University, East Lansing, Michigan 48824

TOMAS C. SHEN (245), Department of Polymer Science, University of Southern Mississippi, Hattiesburg, Mississippi 39401

E. N. SMITH (199), NASA Langley Research Center, Hampton, Virginia 23665

JOSÉ M. SOSA (89), Chemistry Department, New Mexico State University, Box 3C, Las Cruces, New Mexico 88003

G. ALLAN STAHL (259), Department of Chemistry, University of Alabama, Box H, University, Alabama 35486

HOWARD W. STARKWEATHER, JR. (21), Plastic Products and Resins Department, E. I. duPont de Nemours and Company, Experimental Station, Wilmington, Delaware 19898

A. K. ST. CLAIR (199), NASA Langley Research Center, Hampton, Virginia 23665

T. L. ST. CLAIR (199), NASA Langley Research Center, Hampton, Virginia 23665

J. S. TAN (149), Research Laboratories, Eastman Kodak Company, Rochester, New York 14650

Y. O. TU (11), IBM Research Laboratory, Monterey and Cottle Roads, San Jose, California 95193

CARL E. VOIGT (225), DOW Corning Corporation, Corporate Research, Mail #66, Midland, Michigan 48640

C.-P. WONG (71), Department of Chemistry, Carnegie-Mellon University, Pittsburgh, Pennsylvania 15213

HUBERT A. WOOD (99), Department of Chemical Engineering, University of Houston, Houston, Texas 77004

PREFACE

The papers contained in this book were presented in the symposium on "Structure-Solubility Relationships in Polymers" held by the American Chemical Society at its 172nd Meeting in San Francisco, August 30-September 3, 1976. The final manuscripts were collected at the meeting so that the authors could expand considerably on the brief abstracts that appeared in *Polymer Preprints* and could incorporate their most recent results. The joint sponsorship of this symposium by the Division of Polymer Chemistry and the Division of Organic Coatings and Plastics Chemistry in this centennial year for the Society was most appropriate. Polymer solutions have long been a major concern of the Coatings and Plastics Division due to their large consumption by the coatings industry, while theoretical and fundamental aspects of polymer solubility have appeared frequently in *Polymer Preprints*.

In organizing this symposium, we attempted to blend reports on the latest in solubility theory with experimental papers on the effects of structural modifications on solvent interactions. In fact, many of the contributed papers correlate modern solubility theory with experimental results. While seeking a nearly equal amount of industrial and academic participation, we also tried to cover a broad spectrum of polymer types. We feel that the result of these efforts is a comprehensive compilation of the recent significant work in polymer solubility.

We believe that the papers contained herein by the foremost authorities in the field are not only extremely informative but will also form the basis for considerable future research. Hence, we express our sincere appreciation to the contributors for making the symposium a great success and for making the publication of this volume possible.

OPERATIONS ON SWOLLEN THEORIES WITH OCCAM'S RAZOR

Joel H. Hildebrand
Department of Chemistry
University of California
Berkeley, California 94720

This is an opening address invited by chairman Drs. Seymour and Harris presumably in recognition of the author's contribution to solubility theories. This constitutes a critique of theories of liquids and transport phenomena that violate a principle stated by Alfred North Whitehead (Principle of Relativity, Cambridge University Press, 1922):

> The philosophy of science is the endeavor to formulate the most general characters of things observed. These sought-for characters are to be no fancy characters of a fairy tale enacted behind the scenes. They must be observed characters of things observed.

OPERATIONS ON SWOLLEN THEORIES WITH OCCAM'S RAZOR

Joel H. Hildebrand
University of California

William of Ockham, Surrey, England, born about 1300 A.D., became known as "the invincible doctor" because of his cutting criticism of ecclesiastical dogma. One of his utterances became famous as "Occam's Razor." In the original Latin it is:

Essentia non sunt multiplicanda praeter necessitatem

I translate it rather freely as: *The theory of a phenomenon must not be more complicated than is necessary*.

I employ this instrument to operate upon published explanations of phenomena in simple liquids. I know of no subject that has been too overloaded with complications.

My first illustration is the "cage model" of diffusion depicted in the book "The Solubility of Nonelectrolytes" by Hildebrand and Scott (1). A molecule is represented as imprisoned in a cage of molecules, bumping about with a certain frequency. Suddenly it acquires, from an unexplained source, an "activation energy" sufficient to enable it to burst out of its cage into a "hole" in a leap of one diameter. The unreality of that description of diffusion is explained in "Regular and Related Solutions" by Hildebrand, Prausnitz and Scott (2).

> Alder and Einwohner calculated mean free paths in dense fluids from a study of molecular dynamics of 10^6 free paths in a system of 500 hard spheres. . . . The fluid free path distributions. . . are without a peak near a particle diameter or . . . at any other distance. . . . At the highest density considered, $v/v_o = 1.6$, the probability of a free path longer than a hard sphere diameter is 4×10^{-8}.

In another paper, Alder and Hildebrand (3), in a brief communication wrote:

> The method most commonly used for representing the temperature dependence of viscosity and diffusivity of liquids is to plot their logarithms against $1/T$, by analogy with the Arrhenius equation for chemical rate constants, and interpreting the slopes of the

> lines thus obtained as "activation energy." This designation implies the presence of barriers against freedom of flow, imagined as consisting of some sort of quasi-lattice structure. We wish to call attention to evidence that no activation is involved in these processes, and that their variations with temperature are quite accurately represented by nonexponential equations. . . .
>
> All molecules participate equally in the thermal agitation that produces random disorder. . . . Vacancies that may exist in a crystal become randomly distributed in liquid as intermolecular space. . . . No directive force. . . acts upon molecules of a nonpolar liquid. They diffuse simply because thermal motions keep them ever on the move.

Evidence of maximum disorder can be seen in the entropy of vaporization of nonpolar liquids, illustrated in Fig. 1.

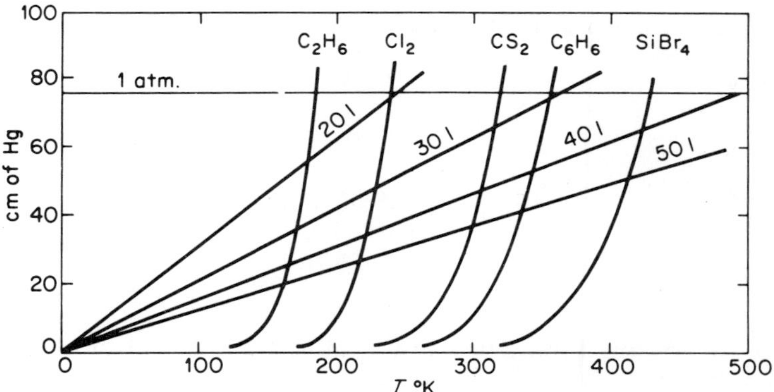

Fig. 1.

Vapor pressures below one atmosphere are plotted againt $\underline{T}$ for a number of liquids. The straight lines represent different volumes. The tangents at the intercepts multiplied by $\underline{R}$ give values of entropy of vaporization; they are equal to the same volume of vapor. It is a kind of "corresponding states" relation. It has been called "the Hildebrand rule." It replaces the "Trouten rule," which compares entropies at one atmosphere, where slopes are not the same. Associated liquids give steeper vapor pressure curves and larger values of entropy of vaporization; therefore agreement with the Hildebrand rule indicates maximum freedom of molecular motions in the liquid.

Hermsen and Prausnitz (4) calculated values of entropy of vaporization to 49.5 liters of 20 alkanes whose boiling points range from 92.6K to 611.9K. The values found were all 22.3 ± 0.1 cal. deg^{-1} mol^{-1}. This finding rules out all structured models, with their "cages," cells," clusters" and "liquid lattices." It permits statistical calculation of relative numbers of contacts of like and unlike species of molecules in a mixture. It indicates that simple liquids have no structure. Crystal structure completely vanishes upon melting; infinitessimal stress suffices to cause a liquid to flow.

The thermodynamic "equation of state" of simple liquids is:

$$P + (\delta E/\delta V)_T = T(\delta S/\delta V)_T = T(\delta P/\delta T)_V$$

It is approximated by the van der Waals equation where $(\delta E/\delta V)_T$ is a/V^2. The exponent 2 is approximately valid for the rare gases, but increases with molecular complexity as illustrated in Table 1.

TABLE 1

	n in $\delta E/\delta V = V^n$
CCl_4	2.09
$C_2Cl_3F_3$	2.19
$c\text{-}C_4Cl_2F_6$	2.35
$c\text{-}C_6F_{11}CF_3$	2.44

P increases linearly with T at constant volume over long ranges, hence $(\delta^2P/\delta T^2)_V = 0$ and $(\delta C_V/\delta V)_T = 0$

In passing along a continuous van der Waals path from gas to liquid above the critical temperature, at no point does any sort of structure enter the fluid.

My next sample of complexity far in excess of necessity is an equation for calculating the viscosity of a simple liquid:

$$\eta = \frac{Nh}{V} \frac{1}{(1 - e^{-\partial/T})} \frac{6}{\eta k} \frac{V}{(V-V_3)} \exp \frac{a'E_3V_3}{(V-V_3)RT} + \frac{V - V_3}{V} \frac{(2mkT)^{1/2}}{3\Pi^{3/2}d^2}$$

I give no reference because this is a critique of procedures, not persons.

My chief purpose in research has been to understand natural phenomena. Max Born well described it when he wrote:

> The problem of physics is how the actual phenomena, as observed with the help of our sense organs aided by instruments, can be reduced to simple notions which are suited for precise measurement and used for the formulation of quantitative laws.

As I contemplated "actual" data for the viscosity of a simple liquid such as propane as it changes with temperature, the "simple notion" occurred to me that the flow of a liquid produced by applying vector momentum should be proportional to the free space between the molecules. Expressed in the shorthand of algebra this is

$$\Phi = \underline{B}\ (\underline{V} - \underline{V}_o)/\underline{V}_o = 1/\eta$$

This remains linear for simple liquids to high pressures (544 atmospheres for propane, and up to molar volumes well above boiling point volumes.) Divergence begins as mean free paths become long enough for molecules to acquire random thermal momenta sufficient to oppose the applied vector momentum (Fig. 2).

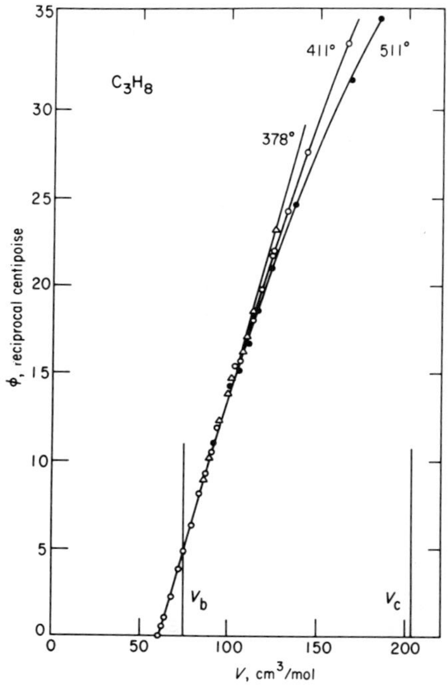

Fig. 2.

Values of the parameter B are related to capacity to dampen vector momentum by transforming it into energy of vibration and rotation.

We have recently applied this interpretation of viscosity to liquid metals, with the result shown in Fig. 3.

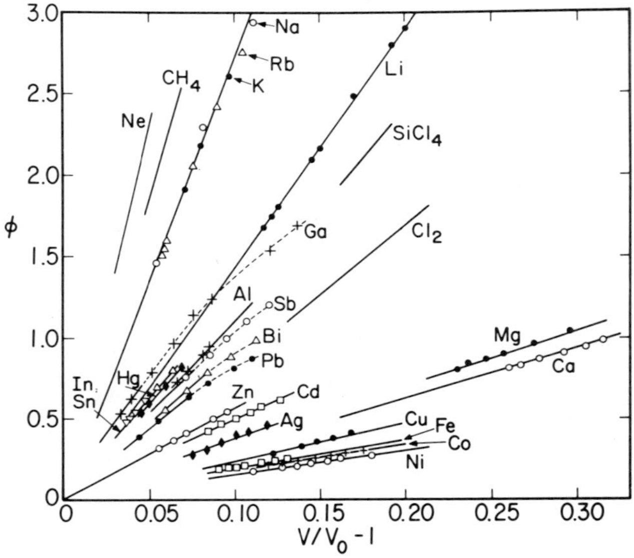

Fig. 3.

The values of $\underline{V_0}$ obtained from plots of ϕ against $\underline{V}$, agree within experimental uncertainty with values of crystal molar volumes obtained from densities at 20° except for Ga, Bi and Sb, which expand upon freezing (5). Plato asserted that the heavenly bodies move in circles because the circle is the most perfect geometrical figure. He was wrong; the most perfect construction is a straight line connecting your theory with facts.

In the case of liquid metals the dampening of vector momentum results from the quasi-chemical bonding of valence electrons of one atom with vacant orbitals of another. It is strongest with the d-electron of the transition elements. Fig. 4 shows viscosities at expansion $\underline{V/V_0} = 1.10$ plotted against cohesion measured by energy of vaporization per cubic centimeter $\Delta E/\underline{V}cm^3$, kcal cm^{-3}, weakest in the alkali metals, increasingly stronger in the transition metals copper to nickel.

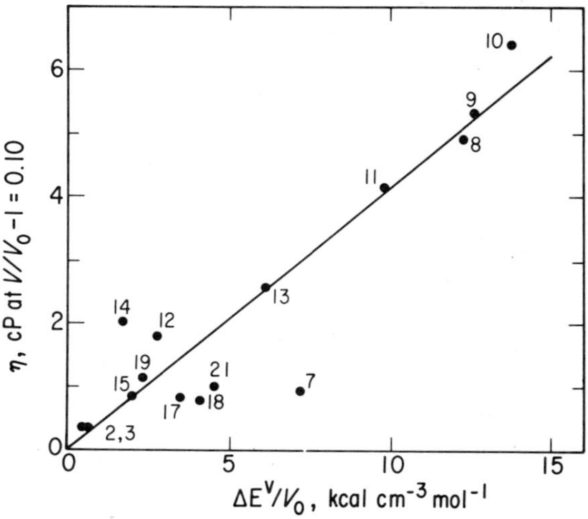

Fig. 4

I end this critique of research methodology with two additional comments. One is that it is often rewarding to plot data in a way that reveals the entropy of a process. I learned a good deal about solubility when I began to plot it as ln x, which multiplied by R is entropy increasing spontaneously toward the maximum permitted by the imposed constraints and is the maximum disorder caused by thermal agitation, against ln T. In a solution of a polymer, it depends upon the freedom of segments to participate in thermal motion, not their positions in imaginary, non-existent latices.

I illustrate the value of acquiring familiarity with the physical meaning of entropy by its role in diffusion. Fig. 5 shows the diffusivity of a number of gases in carbon tetrachloride over ranges of temperature (6). The slope of the lines depend partly upon the cross sections of the gas molecules. Diffusion represents entropy increasing spontaneously toward the limit permitted by the boundary of the system. Fig. 6 shows the entropy of solution of a series of gases in carbon tetrachloride plotted against -Rlnx, which is entropy of dilution. The total entropy is much larger because, as attractive forces decrease in ascending the series, the surrounding solvent molecules gain freedom motion and partial molal volumes become increasingly larger than their intrinsic volumes, with corresponding additions to entropy. Fig. 7 shows that $DV_o^{2/3}$ is a linear function of the entropy of expansion.

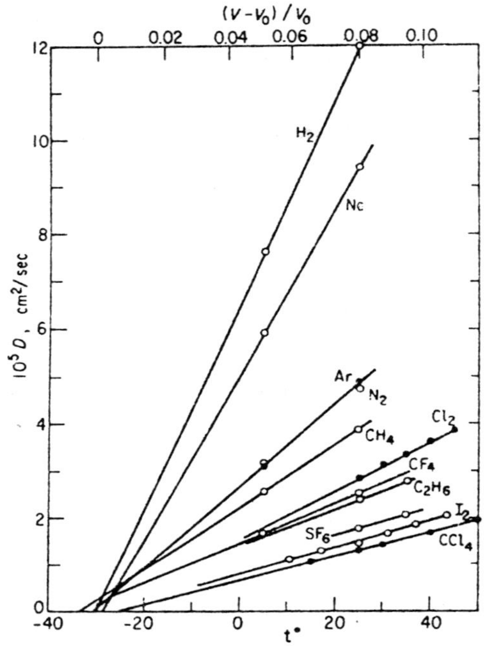

Fig. 5 Variation of diffusivity in centipoise of gases in CCl_4 with temperature and fractional expansion of CCl_4.

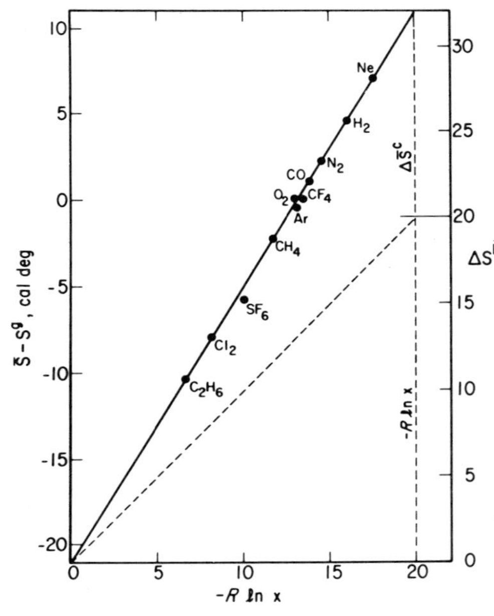

Fig. 6

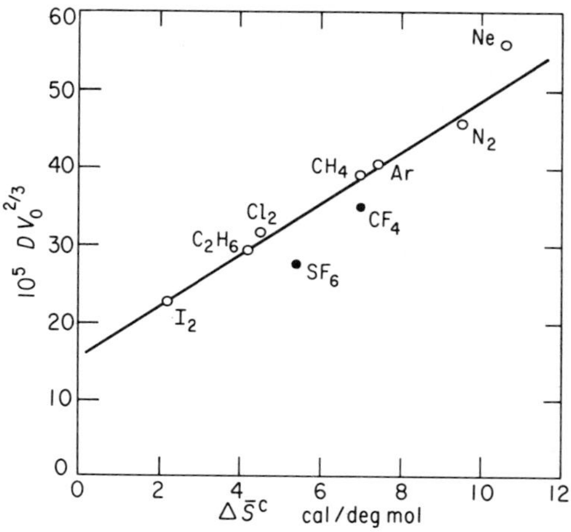

Fig. 7. Relation of diffusivity to molecular cross-section and to configuration entropy.

Finally, I quote a bit of the wisdom of Alfred North Whitehead:

> There can be no true physical science which looks first to mathematics for the provision of a conceptual model. Such a procedure is to repeat the errors of the logicians of the middle ages.

I have heard Peter Debye, confronting a blackboard filled with symbols, ask "What does it mean physically?" Professor of mathematics J. Schwartz, addressing methematicians on "The Pernicious Effect of Mathematics upon Physics?" spoke of "its willingness, like that of a computing machine, to elaborate upon any idea, however absurd. . ." My reaction to a certain paper containing over one hundred mathematical equations was like that of a poet, who wrote:

> As I was laying on the green, a small English book I seen; Carlyle's Essay on Burns was the edition; I left it laying in the same position.

REFERENCES

1. Hildebrand, J.H., and Scott, R. , "The Solubility of Nonelectrolytes," p. 75. Reinhold Publication, New York, 1950.

2. Hildebrand, J.H., Prausnitz, J.M., and Scott, R. "Regular and Related Solutions," Van Nostrand, Reinhold, New York, 1970.
3. Alder, B.J., and Hildebrand, J.H., *I&EC Fundamentals*, 12, 387 (1973).
4. Hermsen, W., and Prausnitz, J.M., *J. Chem. Phys.*, 34, 108 (1961).
5. Hildebrand, J.H., and Lamoreaux, R.H., *Proc. Nat. Acad. Sci.*, April (1976).
6. Hildebrand, J.H., and Lamoreaux, R.H., *Proc. Nat. Acad. Sci.*, 71, 332 (1974).

DYNAMICS OF POLYMER DISSOLUTION

A. C. Ouano, Y. O. Tu, and J. A. Carothers
IBM Research Laboratory

I. INTRODUCTION

The process of mass transport in polymers has been of considerable interest to the industrial and the scientific communities for several decades. In particular, the permeation of gaseous molecules (1-5) and sorption of liquids (6-9) into the polymer matrix have been well studied. However, these activities have rarely been extended to include the dissolution process of polymers and with the exception of the studies of Ueberreiter and his coworkers (10-12), little information is available in the open literature. In this work we made use of some of the well-known properties of glassy and swollen polymers to study the kinematics of polymer dissolution utilizing a phenomenological model to represent the transport processes in the different phases of a dissolving polymer. Unlike the dissolution of low molecular weight crystalline substances such as sucrose in water, where there is no significant permeation of the solvent into the crystalline solid phase, high molecular weight glassy polymers usually exhibit extensive swelling on dissolution. In good solvents glassy polymers undergo a more gradual (relative to the passage of the sucrose crystal into solution in H_2O) transition characterized by sharp changes in modulus or viscosity from a hard glassy substance to an elastic swollen mass, and finally to a low viscosity solution.

II. THE MODEL

As illustrated in the schematics of the dissolution process of glassy polymers shown in Figures 1 and 1a, there are two distinct and sharp changes in the modulus and the diffusion coefficient of the solvent in the polymer with concentration. The first rapid decrease in the modulus G^0 takes place at C_G (see Figure 1), a material parameter which

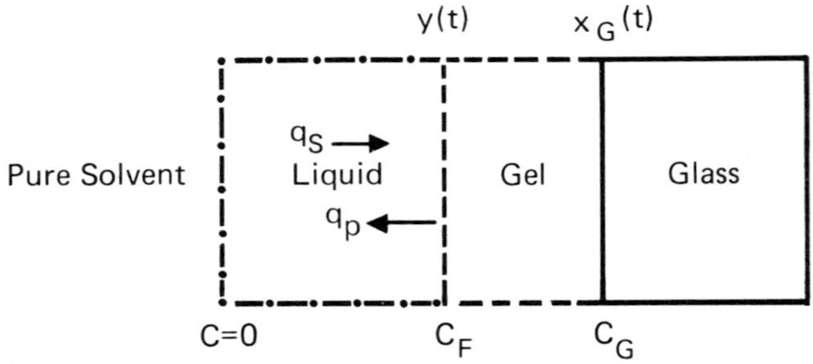

Fig. 1. Schematic of the Phases of Dissolving Polymer

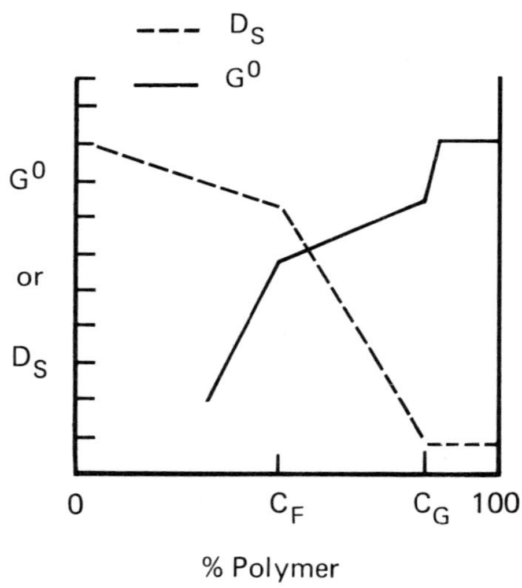

Fig. 1a. Schematic Diagram of D_S and G^0 vs. Conc. D_S - Diff. Coeff. of solvent in polymer, G^0 - modulus

depends only on the polymer/solvent pair. This gel/glass interface is also characterized by a sharp rise in the diffusion coefficient of the solvent in the polymer matrix. The spatial location of this boundary is $X_G(t)$ as indicated in Figure 1. The second sharp drop in the G^0, characterized by a polymer concentration C_F, is the transition from a gel-like substance to a low viscosity solution. For a given polymer/solvent pair and polymer molecular weight, M, C_F is defined quantitatively (13,14). Along with the sharp change in G^0 an abrupt change also takes place at C_F in the slope of the concentration profile. The spatial location of C_F is Y(t) as also indicated in Figure 1. The sharp drop in G^0 at Y(t) has a special practical significance, i.e., if a dissolving polymer is taken out of contact with the solvent, the partition between the dissolved and the undissolved polymer will likely occur at Y(t). Hence, the time derivative of Y(t) is the dissolution rate of the polymer. Thus, the dissolution process of polymers as modeled is described by two diffusion processes separated at Y(t). In the first, the solvent permeates into the polymer causing swelling and eventual "disentanglement" or disassociation of the polymer molecules. In the other, the disassociated polymer molecules diffuse into the solution at a rate that is limited by either the disassociation rate, R (a material parameter depending on the polymer/solvent pair), or by the diffusion capability at C_F to remove the disassociated polymer molecules if R exceeds the latter.

A detailed exposition and a numerical analysis of a one dimensional moving boundary value problem representing the dissolution kinematics of glassy polymers will be described elsewhere (15). Only a summary of the mathematical formulation is presented as follows:

Let the local diffusion coefficients for the solvent and for the dissolved polymer be given by $D_S f_S(c)$ and $D_p f_p(c)$, respectively, where D_S, D_p are the nominal diffusion coefficients and $f_S(c), f_p(c)$ are their concentration dependent factors. Hp is the original polymer thickness, and B is the boundary layer thickness. The terms x and t are taken as the space and time variables, respectively.

Using these notations, the boundary value problem is thus formulated:

(1) The diffusion of solvent in the polymer, $Y^+(t)<x<Hp$

$$\frac{\partial c}{\partial t} = D_s \frac{\partial}{\partial x}\left[f_s(c)\, c\, \frac{\partial c}{\partial x}\right], \quad y^+(t)<x<Hp$$

$$c = c_F, \qquad \text{at } x = Y^+(t), \quad t>0,$$

$$\frac{\partial c}{\partial x} = 0, \qquad \text{at } x = Hp, \qquad t>0 .$$

(2) The diffusion of dissolved polymer, $x<Y^-(t)$.

$$\frac{\partial c}{\partial t} + \frac{dy}{dt}\frac{\partial c}{\partial x} = D_p \frac{\partial}{\partial x}\left[f_p(c)\, \frac{\partial c}{\partial x}\right], \quad x<Y^-(t) .$$

$$c = 0, \qquad \text{at } x = y^-(t) - B.$$

(i) source limited (disassociation rate controlled)

$$D_p f_p(c) \frac{\partial c}{\partial x} = R, \text{ at } x = Y^-(t), \ c<c_F \text{ at } x = Y^-(t).$$

(ii) flux limited (diffusion controlled)

$$c = c_F, \qquad \text{at } x = Y^-(t).$$

(3) Free boundary condition

$$\frac{dy}{dt} = D_p f_p(c) \left.\frac{\partial c}{\partial x}\right|_{x = Y^-(t)} - D_s \left.\frac{\partial c}{\partial x}\right|_{x = Y^+(t)} .$$

Equations (1), (2), and (3) constitute a non-linear diffusion problem known as the Stefan problem (16), the integration of which has been carried out numerically. The numerical treatment of this problem will also be described in detail (15).

Figure (2) shows the concentration profile calculated from the model for a polystyrene with M=180,000 dissolving in methyl ethyl ketone (MEK). The value of C_F was estimated to be 0.15 from the relationship $C_F M=27,000$ (13 and 14). D_s and $f_s(c)$ were estimated from the diffusion coefficient versus c plot shown in Figure (3) which was taken from Asmussen and Ueberreiter's data (10). D_p was obtained from the M versus D_p plot shown in Figure 4. The variations of D_p with concentration was found to be relatively small (17) in the concentration range of interest, as shown in Figure 5. Thus $f_p(c)$ was assumed to be unity. Of all the material parameters used in this study, R is the least understood. Intuitively, its molecular origin should be the same as that of diffusion and viscosity, that is it depends on the segmental mobility of the polymer and the free energy of mixing. However, we are currently unaware of any theory which allows the estimation of R from the molecular or bulk properties of the polymer/solvent pairs. Experimentally, R can be estimated from the solubility rate dependence on solvent flow rate across the surface of the dissolving polymer. At high solvent flow rate the dissolution rate should asymptotically approach R as the boundary layer thickness approaches zero. In this condition, the diffusion of disassociated polymer in solution is controlled primarily by the dissaociation rate R.

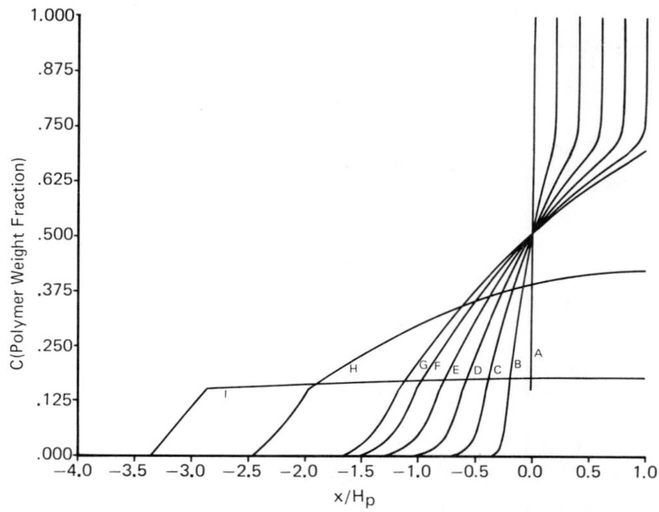

Fig. 2. Computer simulation of the time lapse concentration profile of NBS-705 polystyrene dissolving in MEK. A to I are increasing contact time with the solvent. $H_p = 10\text{-}4$cm, $B = 10^{-2}$cm, $C_F = 0.15$, $C_G = 0.7$, $D_S = 3{\times}10^{-6}$cm^2/sec, $R = 5{\times}10^{-4}$ cm/sec.

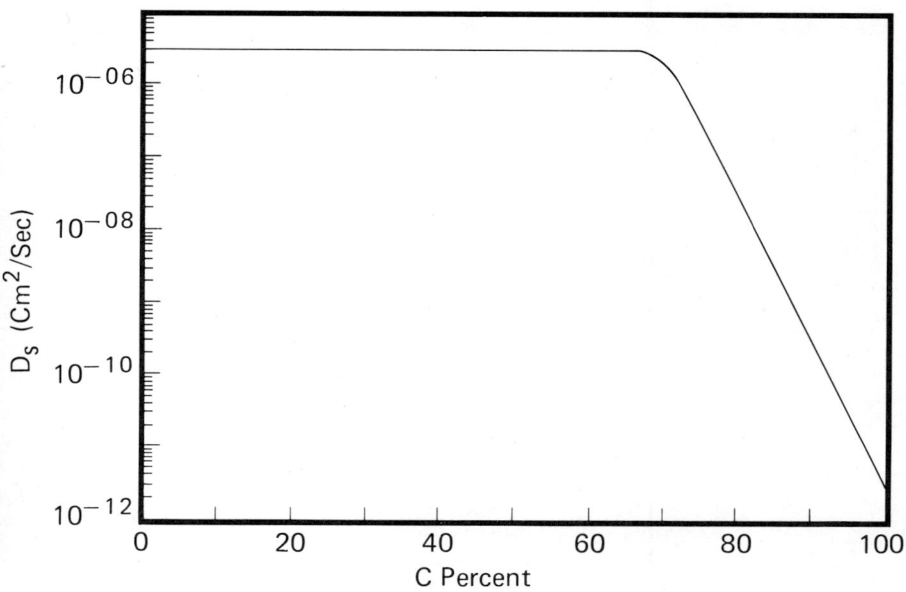

Fig. 3. Diffusion coefficient (D_S) of MEK in polystyrene plotted against the polymer concentration.

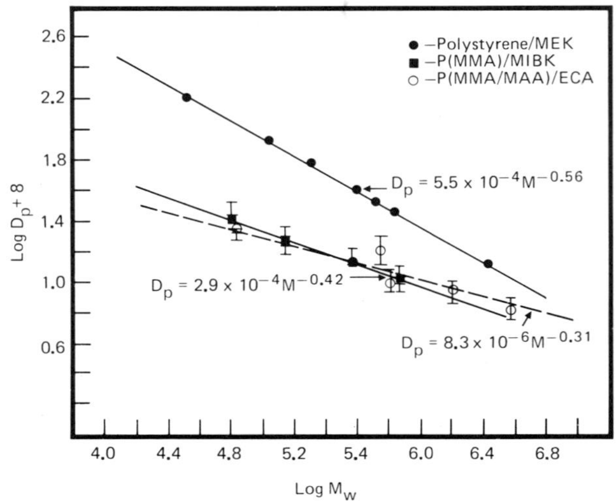

Fig. 4. Diffusion coefficient (D_p) of some polymers in various solvents plotted against their molecular weight (M).

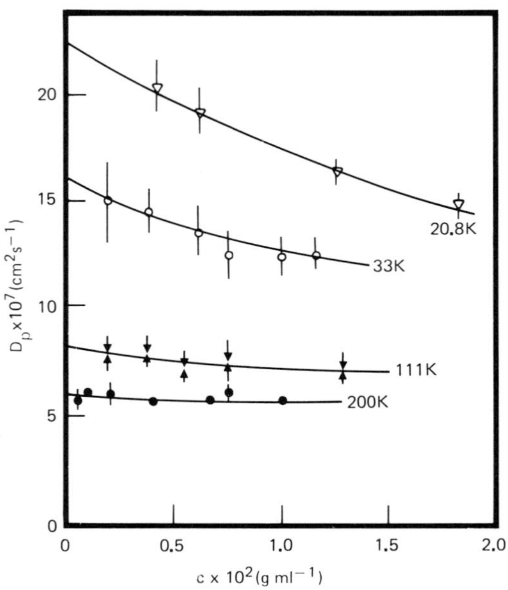

Fig. 5. Diffusion coefficient of polystyrene in MEK plotted against the polymer concentration. (Data is taken from T. A. King et al., (17).) (by permission of the publishers, IPC Business Press Ltd.©)

Curves A to I in Figure 2 are simulated time evolution of the concentration profile of the dissolving polymer. Curve A is the concentration profile at the instant of polymer contact with the solvent and curve I is after the solvent has thoroughly swollen the polymer. The inflection points at c=0.15 and c=0.70 represent the liquid/gel and the gel/glassy boundaries. The concentration profiles in Figure 2 depict a dissolution process where the initial solvent flux into the polymer matrix is higher than the flux of the polymer into the solvent. This results in the complete swelling of the polymer as illustrated by curve I. The concentration profiles in Figure 2 also depict a polymer which is dissolving under a flux limited or diffusion controlled condition ($R>D_p\,\partial c/\partial x$), characterized by the rapid approach to C_F of the polymer concentration in the liquid side of the liquid/gel boundary.

III. EXPERIMENTAL

To follow the rate of dissolution and the motions of the different boundaries, Y(t) and $X_G(t)$, a critical angle illumination microscopy technique (this provides good contrast between boundaries) which allows an in situ dissolution rate measurement was developed. The apparatus consists of a flow through cell in which the polymer sample is bounded by a quartz cell window on two sides for microscopic observation and a Teflon spacer bounding the other three sides. Thus in this configuration only one surface of the sample is exposed to the solvent, simulating a one dimensional dissolution process. The sample cell was mounted on a microscope X-Y stage driven by a set of micrometers, permitting the measurement of Y(t) and $X_G(t)$ to an accuracy of 2.5μm. A television camera was also fitted on the microscope which allowed video taping of the dissolution process in real time as well as in a time lapse mode. Solvent was pumped through the cell by a Waters Associates M-6000 metering pump at a controlled rate. The solvent used was analytical grade methyl ethyl ketone. A manuscript which describes in detail the apparatus and the experimental technique used for the in situ dissolution rate measurement is presently being prepared, and will be published shortly.

IV. DISCUSSION

Figures 6a, 6b, and 6c show front view photomicrographs of a polystyrene (NBS-705; M=180,000) in contact with MEK for various lengths of time. The photomicrograph A was taken a few seconds after the sample came into contact with MEK. After about 200 seconds of contact with MEK, considerable amount of swelling is evident in photomicrograph B. The gel layer was found to continue to grow as shown in photomicrograph C. Eventually it swells thoroughly before complete dissolution takes place.

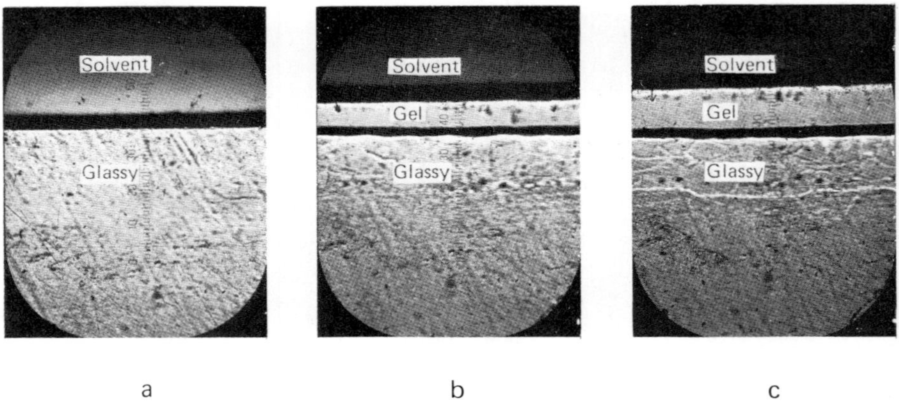

Fig. 6. a, b, and c are photomicrographs of NBS-705 after a few seconds, 200 seconds and about 1000 seconds contact with MEK solvent respectively.

It is evident from the series of photomicrographs shown in Figures 6a, 6b, and 6c that two boundaries are clearly present in polystyrene dissolving in MEK. The positions of these boundaries $Y(t)$ and $X_G(t)$ were followed with time using the X-Y stage and the cross-hairs of the microscope. Figure 7 shows a plot of the position of the liquid/gel ($Y(t)$) and the gel/glass ($X_G(t)$) boundaries (obtained from both the model and have experimental results) against the $\sqrt{t}$. The agreement between the experimental and the computer simulation values of $Y(t)$ and $X_G(t)$ is surprisingly good as evidenced by the data in Figure 7. This good agreement probably stems from the fact that the uncertainty of the value of R did not have a big influence on the simulated values of $Y(t)$. The dissolution rate in this particular example is mainly flux limited.

V. SUMMARY

The dynamics of polymer dissolution (in particular the one-dimensional space kinematics of the solution/gel and the

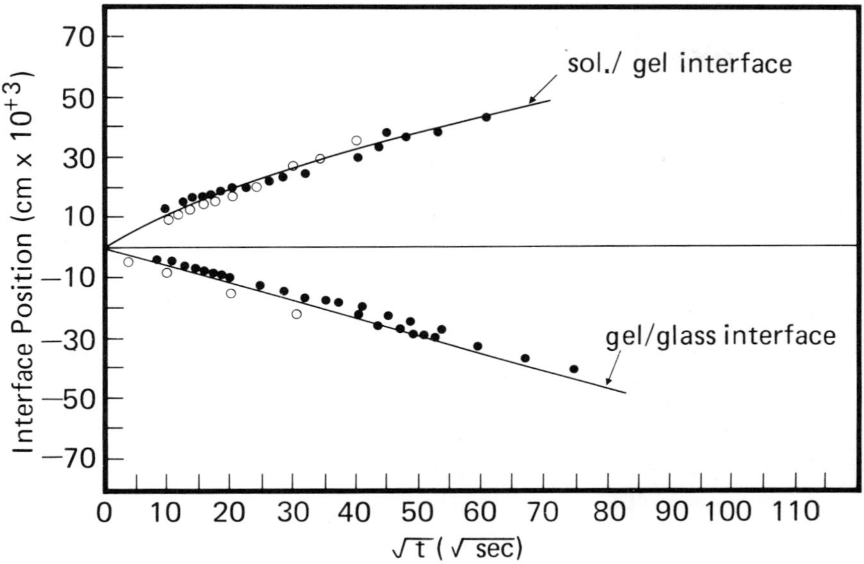

Fig. 7. Positions of the sol./gel and gel/glassy interfaces plotted against the square root of time. ○ *- calculated from the model,* ● *- experimental points,* $H_P = 1$ *cm,* $B = 0.1$ *cm,* $C_F = 0.15$*,* $C_G = 0.7$*,* $D_S = 3\times10^{-6}$ cm^2/sec*,* $D_p = 2\times10^{-7}$ cm^{-2}/sec*,* $R = 5\times10^{-3}$ *cm/sec.*

gel/glassy interfaces) can be adequately predicted by a phenomenological model. This model describes the transport processes in the different phases of a dissolving polymer. The important material parameters required in the model are: (a) D_s and D_p - the nominal diffusion coefficients of the solvent in the solid phases and of the polymer in the liquid phase, respectively; (b) $f_s(c)$ and $f_p(c)$ - the concentration dependence factors for the diffusion coefficients of the solvent in the polymer and polymer in the solvent, respectively; (c) C_F and C_G - the polymer concentration at the liquid/gel and at the gel/glassy interfaces, respectively, and (d) R - the disassociation rate of the polymer molecules from the "entangled" rubbery phase into the "free" solute of the liquid phase. The geometrical parameters required are: (a) Hp - the initial thickness of the dissolving polymer, and (b) B - the thickness of the boundary layer in the liquid phase. Thus, this model could be used to investigate the relative importance of the different parameters on the degree of swelling (the relative motions of the Y(t) and $X_G(t)$) and the rate of dissolution.

It is interesting and significant to note that the model predicts the Fickian behavior of the motion of $X_G(t)$. The model as well as the experimental result show the linear dependence of the gel/glassy interface location with the $\sqrt{t}$. As for the Case II type (8) dissolution process, work is continuing to extend the model to include nonlinear dependence of $X_G(t)$ on the $\sqrt{t}$.

IV. REFERENCES

1. Frisch, H. L., J. Polym. Sci. B3, 13 (1965).
2. Barrer, R. M., Nature 140, 106 (1937).
3. Kumins, C. A., and Roteman, J., J. Polymer Sci. 55, 683 (1961).
4. Michaels, A. S. and Bixler, H. J., J. Polymer Sci. 50, 393 (1961).
5. Stannett, V., "Simple Gases," in Diffusion in Polymers, (J. Crank and G. Park, Eds.), Academic Press, London and New York, 1968.
6. Fugita, H., Fortschr. Hochpolym. Forsch. 3, 1 (1961).
7. Rogers, C. E., Semancik, J. R., and Keegen, S., "Transport Processes in Polymers," in Structure and Properties of Polymer Films (R. W. Levy and R. S. Stein, Eds.) Plenum Press, New York-London, 1973.
8. Kwei, T. K., and Wang, T., "Diffusion in Glassy Polymers," in Permeability of Plastic Films and Coating (H. B. Hopfenberg, Ed.) Plenum Press, New York, 1974.
9. Crank, J. and Robinson, C., Proc. Roy. Soc. A204, 549 (1950).
10. Ueberreiter, K., "The Solution Process," in Diffusion in Polymers (J. Crank and G. Park, Eds.) Academic Press, New York and London, 1968.
11. Asumussen, F. and Ueberreiter, K., Kolloidzeitschrift 185, 1 (1962).
12. Ueberreiter, K., and Kirchner, P., Makromolek. Chem. 87, 32 (1965).
13. Berry, G. C. and Fox, T. G., Adv. Polymer Sci. 5, 261 (1968).
14. Graessley, W. W., Advances in Polymer Sci. 16, 1 (1974).
15. Tu, Y. O., and Ouano, A. C., to be published, IBM Journal of Research.
16. Cohen, H., "Non-linear Diffusion Problems," in Studies in Applied Math., Vol. 7 (A. H. Taub, Ed.) 1971.
17. King, T. A., Knox, A., and McAdam, J. D., Polymer 14, 293 (1973).

CLUSTERING OF SOLVENTS ABSORBED IN POLYMERS

Howard W. Starkweather, Jr.
Plastic Products and Resins Department
E. I. du Pont de Nemours and Company

I. INTRODUCTION

Following earlier work by Kirkwood and Buff (1), Zimm (2) developed a way to determine the degree of clustering or nonrandom mixing in two-component systems without regard to any preconceived model. In fact, use of this approach can lead to possible models. The treatment was applied to mixtures of polymers and solvents by Zimm and Lundberg (3). It has been helpful in gaining understanding of the relationship among water molecules absorbed in natural and synthetic polymers (4,5).

The mean number of solvent molecules in the neighborhood of a given solvent molecule in excess of the mean concentration of solvent is given by

$$c_1G_{11} = (1 - \phi_1)\left(\frac{\partial \ln \phi_1}{\partial \ln a_1}\right)_{p,T} - 1 \qquad (1)$$

where G_{11} is the cluster integral, and c_1, ϕ_1, and a_1 are the molar concentration, volume fraction, and activity of the solvent. It follows from this expression that the average number of solvent molecules in a cluster is

$$c_1G_{11} + 1 = (1 - \phi_1)\left(\frac{\partial \ln \phi_1}{\partial \ln a_1}\right)_{p,T} \qquad (2)$$

In a mixture of polymer and solvent, some clustering is expected because of their different molecular volumes. Since a chain of polymer segments is a cluster of sorts, there is some tendency toward a complementary segregation of the solvent molecules. This becomes more pronounced as ϕ_1 is increased. If the molecular weight is high and the solution is not too dilute, the Flory-Huggins equation may be written as

$$\ln a_1 = \ln \phi_1 + (1 - \phi_1) + \chi (1 - \phi_1)^2 \quad (3)$$

where χ is the interaction parameter.

$$\left(\frac{\partial \ln a_1}{\partial \ln \phi_1}\right)_{p,T} = 1 - \phi_1 - 2\chi\phi_1(1 - \phi_1) \quad (4)$$

From equations 2 and 4, the cluster size is

$$c_1 G_{11} + 1 = \frac{1}{1 - 2\chi\phi_1} \quad (5)$$

In a Θ solvent where there is no energetic preference between polymer-solvent and solvent-solvent contacts, $\chi = 0.5$ and

$$c_1 G_{11} + 1 = \frac{1}{1 - \phi_1} \quad (6)$$

The average cluster size is larger for a poor solvent and smaller for a good one. Equation 6 provides a standard for judging whether there is more or less clustering of the solvent than would be expected solely from the polymeric nature of the solute. As χ is increased, so is the degree of clustering. In fact, when ϕ_1 reaches $\frac{1}{2\chi}$, the clusters become indefinitely large and phase separation occurs.

If χ varies with ϕ_1,

$$c_1 G_{11} + 1 = \frac{1}{1 - 2\chi\phi_1 + (1 - \phi_1)\left(\frac{\partial \chi}{\partial \ln \phi_1}\right)_{p,T}} \quad (7)$$

This means that there are further increases in clustering if χ increases with increasing concentration of polymer (decreasing ϕ_1). Experimentally, this is usually the case (6).

An expression for the dependence of the interaction parameter on the volume fraction of solute, $\phi_2 = 1 - \phi_1$, has been given by Koningsveld and Kleintjens (7).

$$\chi = \alpha + \frac{\beta(1-\gamma)}{(1-\gamma\phi_2)^2} \quad (8)$$

The coordination number, z, is said to be approximately equal to $2/\gamma$. Combining Equations (7) and (8) with proper differentiation leads to

$$c_1G_{11} + 1 = \frac{1}{1 - 2\alpha\phi_1 - \dfrac{2\beta(1-\gamma)\phi_1}{(1-\gamma\phi_2)^3}} \quad (9)$$

With this bacground, let us consider the various kinds of clustering which can be deduced from published data on a number of polymer-solvent combinations.

II. "FLORY-HUGGINS" CLUSTERING

A system which is known to follow the Flory-Huggins equation fairly well is natural rubber-benzene (8). The data for cluster size vs ϕ_1 for this system are plotted in Figure 1 along with Equation 5 for several values of χ. The relationship for $\chi = 0.4$ is followed fairly well for values of ϕ_1 up to 0.5. At higher levels of solvent, the cluster size is less than was predicted from the Flory-Huggins equation with a constant χ. Eichinger and Flory (9) have concluded that this deviation from the Flory-Huggins equation can be attributed to the different equation-of-state characteristics of the components.

Another system which exhibits "Flory-Huggins" clustering is polystyrene-propyl acetate (10). The data on clustering shown in Figure 1 are in general agreement with the report that $\chi = 0.64$ at 25°C. However, the average cluster size remains between 2.3 and 2.8 for values of ϕ_1 from 0.36 to 0.56. Perhaps, this indicates a special stability for dimers or trimers of propyl acetate in polystyrene.

III. UNCLUSTERED SORPTION OF SOLVENTS

There are some systems where c_1G_{11} is so close to zero over a wide range of composition that the positions of the solvent molecules are essentially uncorrelated with each other. They lack even the clustering which is expected from the polymeric nature of the solute. Among the most striking examples of this class are the mixtures of chloroform

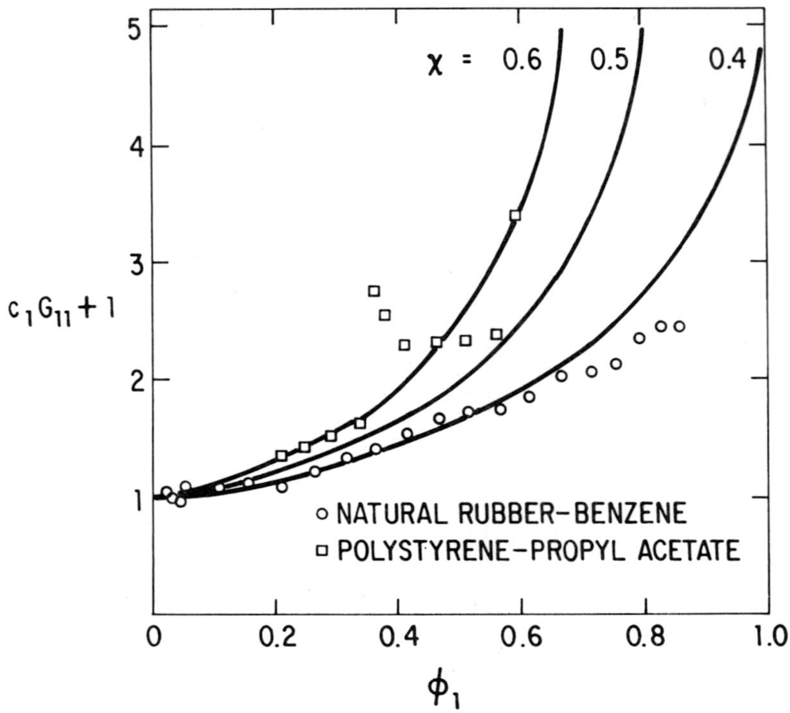

Fig. 1. "Flory-Huggins" Clustering

with polybutadiene (11) or polystyrene (10). As shown in Figure 2, the average cluster size remains less than 1.2 for values of ϕ_1 up to 0.6. In terms of the Flory-Huggins equation, this would be equivalent to a value of χ less than 0.14. At higher levels of chloroform in polystyrene, there does appear to be some clustering reaching an average cluster size of two at $\phi_1 = 0.8$.

Somewhat less extreme examples of almost unclustered sorption are polybutadiene-benzene (12) and polystyrene-toluene (13). The average cluster size remains under 1.5 for values of ϕ_1 up to 0.5. When the level of toluene exceeds one molecule per monomer unit in polystyrene (indicated by the arrow in Figure 2), clustering develops rapidly.

IV. SORPTION ON EXISTING CLUSTERS

It is of interest to consider the number of clusters as well as their size. Let w be the amount of solvent mixed with a given quantity of polymer.

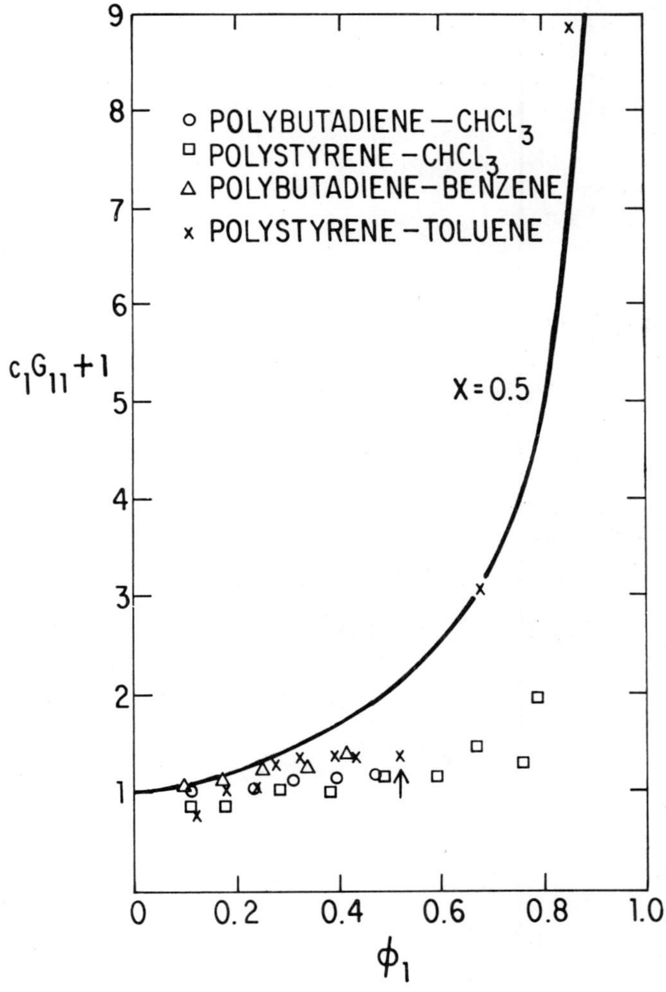

Fig. 2. Sorption with Minimal Clustering

This can be expressed as grams per 100 grams of polymer or solvent molecules per 100 monomer units. The number of clusters in corresponding units is given by

$$\frac{w}{c_1 G_{11} + 1} \tag{10}$$

One way of presenting the data is to plot this quantity against w. In regions of composition where additional solvent is absorbed on existing clusters, the number of solvent molecules increases but the number of clusters does not and the curve will be horizontal. This kind of behavior was found for the sorption of water in serum albumin at high relative humidities (5).

One kind of system where sorption on existing clusters is likely to occur is a combination of a polar solvent and a hydrocarbon polymer. Data for natural rubber with acetone, methyl ethyl ketone (MEK) and ethyl acetate (14) are shown in Figure 3.

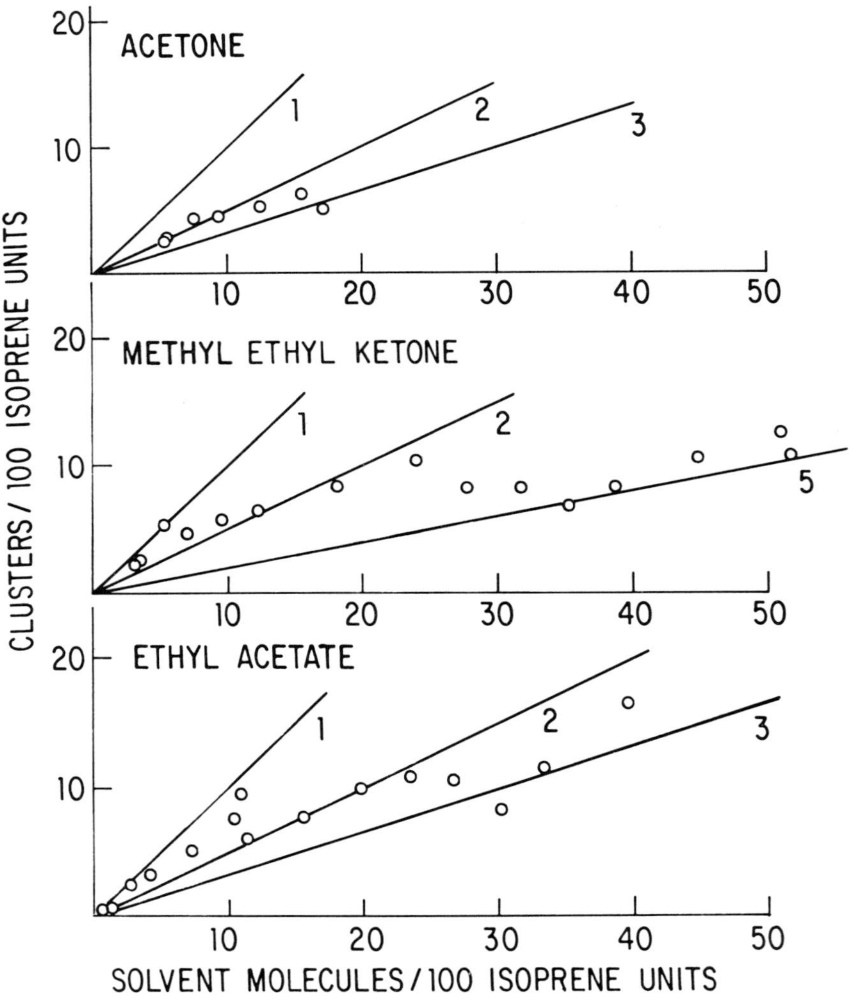

Fig. 3. Clustering of Polar Solvents in Natural Rubber.

The diagonal lines correspond to the indicated cluster sizes and have slopes equal to the reciprocal of cluster size. In each case, the number of clusters increases gradually until the average cluster size is two and then remains constant. The maximum cluster size computed from the experimental data is about three for acetone and ethyl acetate and five for MEK. The maximum number of clusters per 100 isoprene units is about five for acetone and about ten for MEK and ethyl acetate.

Similar behavior is found when acetone (10) or MEK (13) is absorbed in polystyrene. As shown in Figure 4, the number of clusters per 100 styrene units remains constant at about 20 for acetone and 25 for MEK as the average cluster size increases from two to five or six.

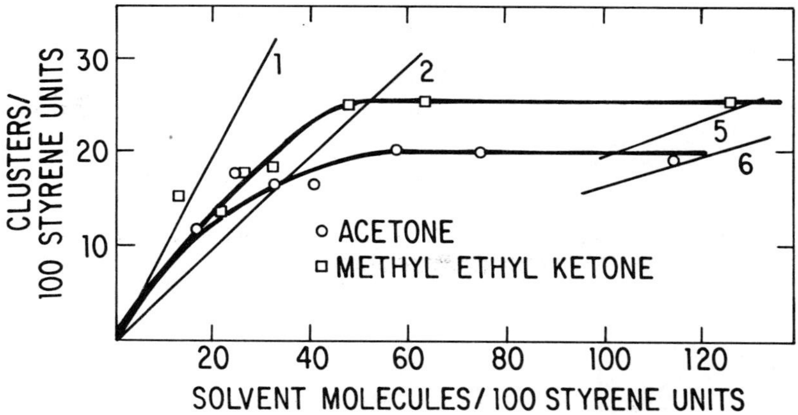

Fig. 4. Clustering of Ketones in Polystyrene.

Sorption by addition to existing clusters is not limited to polar solvents. Data for benzene in polyisobutylene (12,15) are presented in Figure 5. As in the systems discussed above, the number of clusters increases gradually until the average cluster size is about two. The number of clusters then remains near 14 per 100 isobutylene units as the average cluster size increases from two to seven.

The properties of these compositions having a constant number of solvent clusters in a given amount of polymer are summarized in Table 1. The systems all involve relatively poor solvents for the polymers involved. Clearly, solvent dimers play a

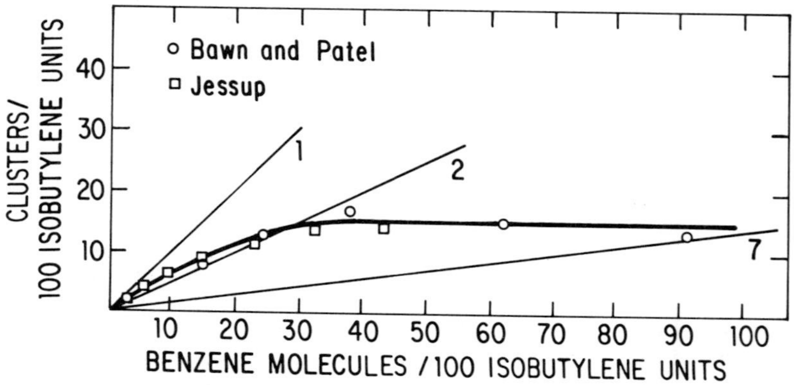

Fig. 5. Clustering of Benzene in Polyisobutylene.

special role in these systems. Some evidence for this has also been seen in Figure 1 for polystyrene-propyl acetate, a system only marginally different from those considered here.

TABLE 1
Compositions of Constant Cluster Number

Polymer	*Solvent*	*Range of Average Number of solvent Molecules per Cluster*	*Monomer Units per Cluster*
Natural rubber	Acetone	2-3	20
Natural rubber	MEK	2-5	10
Natural rubber	Ethyl acetate	2-3	10
Polystyrene	Acetone	2-6	5
Polystyrene	MEK	2-5	4
Polyisobutylene	Benzene	2-7	7

It is of interest to consider the amount of polymer required to package a cluster of solvent molecules. This varies from four to twenty monomer units per cluster. If clusters are separated by a monolayer of polymer segments, the amount of polymer surrounding a cluster would be twice these amounts. In some cases, this seems to be close to the plausible minimum.

V. COALESCENCE OF CLUSTERS

If the sorption of solvent on clusters continues beyond a certain point, the walls of polymer separating the clusters will break. When this happens, the number of clusters will pass through a maximum and decrease. This happens for polystyrene solutions in cyclohexane (16) as shown in Figure 6.

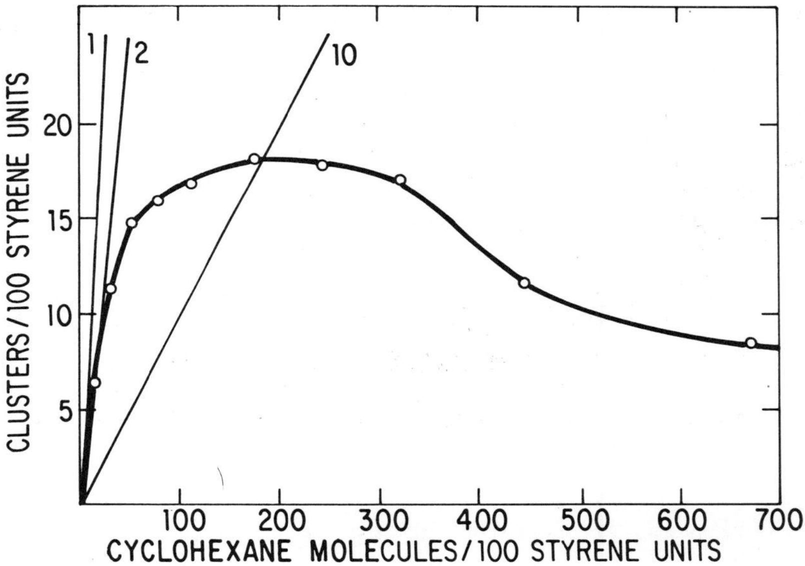

Fig. 6. Clustering of Cyclohexane in Polystyrene.

This is the system for which Equation 8 for the dependence of the interation parameter on composition was developed (7). The number of clusters reaches a maximum value of 18 per 100 styrene units at an average cluster size of 10. The addition of more solvent results in a smaller number of clusters of greater average size.

Another system which exhibits coalescing clusters is polystyrene-toluene (13) (Figure 7). As stated earlier, there is very little clustering up to about one molecule of toluene per styrene unit. Beyond this point, the average cluster size increases up to about nine and the number of clusters decreases.

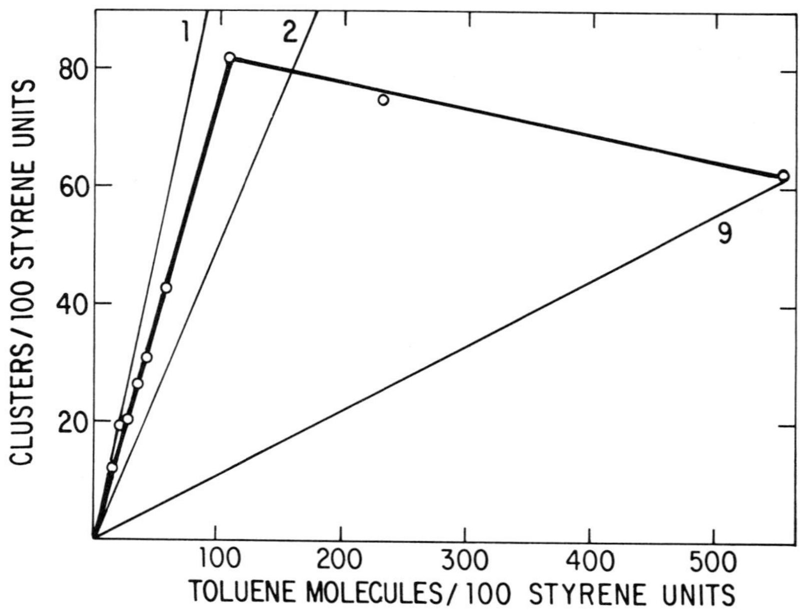

Fig. 7. Clustering of Toluene in Polystyrene.

VI. REFERENCES

1. Kirkwood, J.G., and Buff, F.P., J. Chem. Phys., 19, 774 (1951).
2. Zimm, B.H., J. Chem. Phys., 21, 934 (1953).
3. Zimm, B.H., and Lundberg, J.L., J. Phys. Chem. 60, 425 (1956).
4. Starkweather, H.W., Polymer Letters, 1, 133 (1963).
5. Starkweather, H.W., Macromolecules, 8, 476 (1975).
6. Flory, P.J., Discussions of the Faraday Society, 49, 1 (1970).
7. Koningsveld, R., and Kleintjens, L.A., Macromolecules, 4, 637 (1971).
8. Gee, G., and Treloar, L.R.G., Trans. Faraday Soc. 38, 147 (1942).
9. Eichinger, B.E., and Flory, P.J., Trans.Faraday Soc., 64, 2035 (1968).
10. Bawn, C.E.H., and Wajid, M.A., Trans. Faraday Soc., 52, 1658 (1956).
11. Booth, C., Gee, G., Jones, M.N., and Taylor,W.D., Polymer, 5, 353 (1964).

12. Jessup, R.S., *J. Research National Bureau of Standards*, 60, 47 (1958).
13. Bawn, C.E.H., Freeman, R.F.J., and Kamaliddin, A.R., *Trans. Faraday Soc.*, 46, 677 (1950).
14. Booth, C., Gee, G., Holden, G., and Williamson, G.R., *Polymer* 5, 343 (1964).
15. Bawn, C.E.H., and Patel, R.D., *Trans. Faraday Soc.*, 52, 1664 (1956).,
16. Krigbaum, W.R., and Geymer, D.O., *J. Am. Chem. Soc.*, 81, 1859 (1959).

STRUCTURE - SOLUBILITY PARAMETER RELATIONSHIPS IN ALCOHOLS

Kenneth D. Nisbet
Technical Coatings Inc.

I. INTRODUCTION

The evolution of the three component solubility parameter theory during the past decade has resulted in a strong formulating technique for the polymer industry. The diversified approach of several authors in considering the polar influences in solvents and solvent-polymer interactions has brought forth: experimental techniques to assign values to the solubility parameter components (hereafter abreviated SPC), regression techniques to predict approximate SPC, and insights as to the molecular dynamics involved. See Barton(1) and Hansen and Beerbower(2) for reviews.

For polymer applications the major deficiency in the present techniques for obtaining SPC is the lack of a simple method both to predict the polymer-solvent interactions solely from structural considerations and to confirm the values experimentally. The SPC of multifunctional polymers containing polar groups in close proximity have been intractable when hydrogen bonding was a major mechanism of functional group association. Alcohols are generally considered to be strongly hydrogen bonded liquids. They have even been adequately described as pseudo polymers due to their strong intermolecular bonding. The purpose of this paper is to describe a concept which allows very accurate calculation of the SPC of a series of multifunctional, noncyclical, aliphatic alcohols.

As applied to pure, non-polar solvents, the solubility parameter was defined by Hildebrand and Scott(3) as

$$\delta_{TOT} = \left(\frac{\Delta H_V - RT}{V_m} \right)^{1/2} \qquad (1)$$

where δ_{TOT} = total solubility parameter
H_V = molar latent heat of vaporization
V_m = molar volume of the pure liquid solvent.

This is a working definition and δ_{TOT} can be calculated from measurable physical properties of the solvents.

Practical, empirical values of δ_{TOT} have been obtained generally through the subjective assessment of the degree of swelling of slightly crosslinked polymers by well characterized solvents. In turn, once δ_{TOT} is determined, unknown solvents can be ranked. The most compatible solvent-polymer pairs result in the greatest degree of swelling. However, the single parameter, δ_{TOT} , must be used with caution. A range of partial swelling of polymers or partial solubilizing of oligomers is commonly observed due to the polymer molecular weight distribution, inconsistent chain sequencing, branching, and impurities due to residual reactants or additives.

II. CALCULATIONS

The ability to calculate meaningful solubility parameters for polymers originated when Small(4) described a linear summation over constants assigned to discrete molecular moieties that yielded an approximation to the total solubility parameter of non-polar molecules

$$V_m \delta = \Sigma F_i \tag{2}$$

where F_i = molar attraction constants of individual molecular groups.

The direct application to non-polar polymers was sufficiently accurate to encourage further development of structural methods in the subsequent years. Although Small's table of molar-attraction constants contained values of polar functionalities, he reported poor correlations with measured solubility parameters when hydrogen bonding was prevalent. This point has been confirmed repeatedly as noted below.

Several authors have followed a Small's-type summation in relatively unsuccessful attempts to conceive and correlate various components of the solubility parameter with a set of constants associated with the polar functionalities: Nygard(5) (-OH, with a hydrogen bonding component),Blanks and Prausnitz (6)(all functionalities, with a polar component), and Gardon(7) (-OH, with a polar component). Gardon suggested that the lack of a simple correlation was due to the complexities of molecular association in response to the donor-acceptor properties of the alcohols. He reasoned that the geometrical mean assumption, apparent in Small's approach, possibly broke down. Hoy(8), likewise, depicted the intramolecular ring formations resulting from intramolecular associations in alcohol-ethers.

From precise, physical, measurable parameters Hoy(9) has generated the solubility parameter components of over 800 compounds. These components are referred to as non-polar, polar, and hydrogen bonding. A regression analysis allowed him to assign Small's-type constants to molecular moieties. He then extended this method to predict the SPC of polymers and unknown compounds. The drawback of this approach in polymer applications, where SPC errors may build up rapidly with molecular weight and relative functional positions, is the lack of any reported experimental technique for confirmation of the range of values predicted for the individual SPC of the polymer.

Hansen(10) has deduced, empirically, an alternate set of SPC based on subjective observations of solvent-swollen polymer samples. A three dimensional grid facilitated assignment of relative SPC values to the polar and hydrogen bonding components after the dispersion component was arbitrarily fixed. Obviously, this method can ascertain the range of the individual SPC of a polymer system while accounting for a molecular weight distribution. Hansen conveniently displays this information as a sphere of solubility. Unfortunately, no general method yet reported can quantitatively predict these solvent SPC values.

A marriage of the two techniques would seem to be in order. An empirical technique should be developed to confirm Hoy's method or a regression analysis employed to predict Hansen's values. Hoy is presently working on the former. Published attempts at the latter have so far been unsuccessful. Several principles guided the present work which utilizes a multiple linear regression analysis:

1) Hansen's(2,11) most recently modified data on alcohols were used;
2) Constants were ultimately assigned to polar interactions rather than molecular moieties;
3) The summation was in terms of energy for each SPC to yield

$$V_m\delta_i^{\,2} = \Sigma\ (\text{interactions}) + \text{const} \qquad (3)$$

rather than Small's-type units of $(\text{cal}\cdot\text{cc})^{\frac{1}{2}}$.
4) This structural method was required to be simple to apply and not involve detailed structural information, i.e. neither bond angles, bond lengths, nor preferred orientations.

Recently, Koenhen and Smolders(12) have provided a table of Small's-type dispersion-molar attraction constants based on Hansen's data

$$V_m\delta_d = \Sigma\ F_{id} \qquad (4)$$

where F_{id} = dispersion-molar attraction constants. They concluded that the dispersion constants appear to be additive within experimental error. Table I lists Hansen's modified

TABLE I

Solubility Parameter Components

No.	Alcohol	δ_{TOT}	Dispersion δ_d	Polar δ_p	Hyd. Bond δ_h	Molar Vol. V_m	No. of OH N
1	N-Propanol	11.97	7.78	3.3	8.5	75.2	1
2	N-Butanol	11.30	7.81	2.8	7.7	91.5	1
3	N-Pentanol	10.61	7.81	2.2	6.8	109.0	1
4	2-Ethyl Butanol	10.38	7.70	2.1	6.6	123.2	1
5	Methyl Isobutylcarbinol	9.72	7.47	1.6	6.0	127.2	1
6	Propylene Glycol	14.80	8.24	4.6	11.4	73.6	2
7	Ethylene Glycol	16.30	8.25	5.4	12.7	55.8	2
8	1,3-Butanediol	14.14	8.10	4.9	10.5	89.9	2
9	Glycerol	17.65	8.46	5.9	14.3	73.3	3
10	2-Propanol	11.5	7.7	3.0	8.0	76.8	1
11	2-Butanol	10.8	7.7	2.8	7.1	92.0	1
12	Iso-Butanol	11.1	7.4	2.8	7.8	92.8	1
13	1-Octanol	10.2	8.3	1.6	5.8	157.7	1
14	1-Decanol	10.0	8.6	1.3	4.9	191.8	1

C. Hansen's Modified Data[11] (25°C)

data for the fourteen alcohols chosen for the present examination. Table II lists the constants of Koenhen and Smolders applied to these fourteen alcohols. Their method appears subject to errors where branching, long chains, or multifunctionality arises. They found the polar component (δ_p) and the hydrogen bonding component (δ_h) impossible to calculate via molar attraction constants.

Hansen and Beerbower(2) have listed a set of group contributions to δ_p and δ_h as

$$V_m \delta_p = \Sigma F_{ip} \tag{5}$$

and

$$V_m \delta_h^2 = N(4650) \tag{6}$$

where N = number of hydroxyls per molecule. Table II again shows the dubious results of this limited approach when applied to the chosen alcohol series. Hansen and Beerbower emphasize the need for a consistent set of units for the SPC.

III. RING CONCEPT

If Hansen's separations of energies for the condensed liquid state is valid, and if he has provided an internally consistent set of empirical data, then each of the components should be attainable through consideration of energetically related phenomena:

$$\Delta E_{TOT} = \Delta E_d + \Delta E_p + \Delta E_h \tag{7}$$

This equation has opened new avenues of thought. However, merely accounting for the presence of a polar group yields an inadequate general description of the SPC. On the other hand, intra- and intermolecular polar interactions are a reasonable route to investigate. The normal degrees of freedom available to a polyatomic molecule, the ensuing nuclear shielding effects, and the molecular placement of hydroxyl groups should lend themselves to simple energetic analyses.

The method to be described explicitly accounts for intramolecular spatial distances of hydrogen atoms to the polar hydroxyl group. The polar interactions generally decrease after six atoms along the chain when the oxygen atom is labeled number one. The spatial association of the oxygen atom with any hydrogen atom forms an enclosed pseudo ring structure and gives rise to the nomenclature RING. Convenient tabulations of the hydroxyl interactions with hydrogen atoms occur in RINGS of 3, 4, 5, and 6 for the most part.

The RING method is exemplified by reference to propylene

TABLE II

SPC Calculated by Other Structural Methods

Alcohol No.	Dispersion δ_d Dev.	Dispersion δ_d % Dev.	Polar δ_p Dev.	Polar δ_p % Dev.	Hydrogen Bonding δ_h Dev.	Hydrogen Bonding δ_h % Dev.
1	0.05	0.6	0.0	0.0	0.6	7.0
2	0.03	0.4	0.1	3.6	0.6	7.8
3	0.04	0.5	0.1	4.5	0.3	4.4
4	0.16	2.1	0.1	4.8	0.5	7.6
5	0.06	0.8	0.4	25.0	0.0	0.0
6	0.24	2.9	0.0	0.0	0.2	1.8
7	0.28	3.4	0.7	13.0	0.2	1.6
8	0.00	0.0	1.1	22.4	0.3	2.8
9	0.08	0.9	1.1	20.0	0.5	3.5
10	0.5	6.4	0.3	10.0	0.2	2.5
11	0.2	2.6	0.1	3.6	0.0	0.0
12	0.0	0.0	0.1	3.6	0.7	9.0
13	0.2	2.4	0.0	0.0	0.4	6.9
14	0.5	5.8	0.0	0.0	0.0	0.0
	Koenhen & Smolders[13]		Hansen & Beerbower[2]		Hansen & Beerbower[2]	
	$V_m\delta_d = \Sigma F_i$		$V_m\delta_p = \Sigma F_i$		$V_m\delta_p{}^2 = N\ (4650)$	
	CH (51) CH_2 (139) CH_3 (201) OH (99)		OH (1) = 250 OH (N) = N (170)		N ≡ no. of OH	

glycol to illustrate the tabulation of hydroxyl interactions. The methylyne, methylene, and methyl hydrogens are counted as a group when encountered. The hydrogens on a particular carbon are assumed to be equivalent. Weighing them 1,2, and 3 times this amount, respectively, is unnecessary since the regression constants would just become multiples of 1, 1/2, and 1/3.

The symbol used to describe a specific RING interaction is

$$R_{\alpha}^{\beta}$$

where R = indicates a RING formation,
α = RING SIZES of 3,4,5,6,(7 and 8+) counting an oxygen atom as no. 1,
β = interaction with a particular hydrogen group where 1,2,3 correspond to $\geq CH$, $>CH_2$, $-CH_3$
and H signifies another hydroxyl hydrogen is involved in a mutual interaction.

In addition, the notation

$$C^{\beta}$$

represents a simple summation of the number and type of carbons present where again β = 1,2,3 = $\geq CH$, $>CH_2$, $-CH_3$.

The first RING interaction illustrated in Figure 1a is a three membered RING methylene interaction (R_3^2) completed with an artificial dashed line.

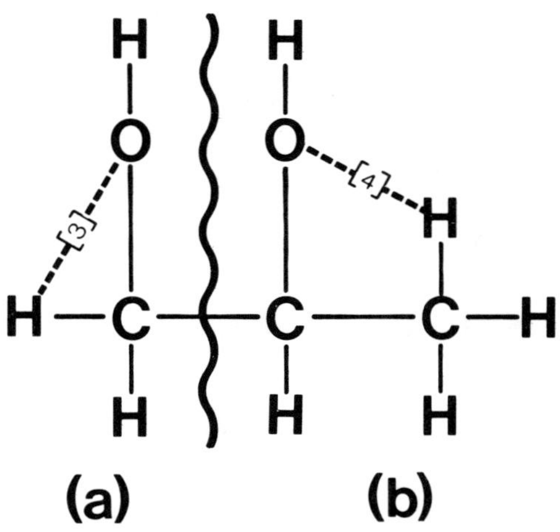

Fig. 1. Depiction of simple RING formation.

Another interaction in pictured in Figure 1b as a four membered RING methyl interaction (R_4^3). All of the possible hydroxyl-hydrogen interactions of propylene glycol are listed in Table III.

Figure 2 illustrates the specific method of counting to form RINGS with reference to any oxygen of a hydroxyl group. This figure also points out that the hydrogens on any carbon are equivalent.

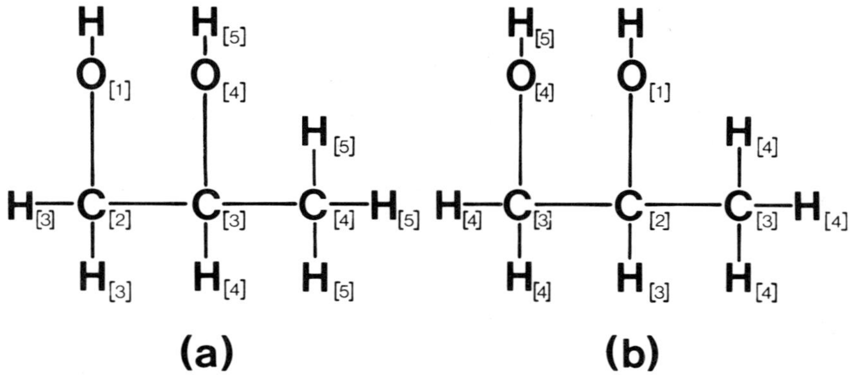

Fig. 2a. Counting from the first oxygen of PG.
2b. Counting from the second oxygen of PG.

The complete interaction matrix for the fourteen alcohols is given in Table IV. For example, the Total for propylene glycol in Table III is shown on line six of Table IV.

IV. RESULTS

A multiple linear regression over the interactions presented in Table IV was performed independently with each set of components from Table I. The statistical t-Test was used to determine which of the 18 interactions were fundamental to each set of components. The statistical F-ratio was used to determine the overall validity of the fit. The SPC results will thus be presented individually.

A. Polar

Polar interactions are considered to arise from dipole-dipole and dipole-induced dipole associations over long dis-

TABLE III

RING Interactions of Propylene Glycol

RING	3		4			5				6		
Group	CH_1	CH_2	CH	CH_2	CH_3	CH	CH_2	CH_3	OH	CH_2	CH_3	OH
Notation	R_3^1	R_3^2	R_4^1	R_4^2	R_4^3	R_5^1	R_5^2	R_5^3	R_5^H	R_6^2	R_6^3	R_6^H
First OH		1	1					1	1			
Second OH	1			1	1				1			
Total	1	1	1	1	1	0	0	1	2	0	0	0

TABLE IV

RING Interaction Matrix

No.	R_3^1	R_3^2	R_4^1	R_4^2	R_4^3	R_5^1	R_5^2	R_5^3	R_5^H	R_6^2	R_6^3	R_6^H	R_7^2	R_{8+}^2	N	C^3	C^2	C^1
1	0	1	0	1	0	0	0	1	0	0	0	0	0	0	1	1	2	0
2	0	1	0	1	0	0	1	0	0	0	1	0	0	0	1	1	3	0
3	0	1	0	1	0	0	1	0	0	1	0	0	0	0	1	1	4	0
4	0	1	1	0	0	0	2	0	0	0	2	0	0	0	1	2	3	1
5	1	0	0	1	1	1	0	0	0	0	2	0	0	0	1	3	1	2
6	1	1	1	1	1	0	0	1	2	0	0	0	0	0	2	1	1	1
7	0	2	0	2	0	0	0	0	2	0	0	0	0	0	2	0	2	0
8	1	1	0	2	1	1	1	0	0	0	1	2	0	0	2	1	2	1
9	1	2	2	2	0	0	2	0	4	0	0	2	0	0	3	0	2	1
10	1	0	0	0	2	0	0	0	0	0	0	0	0	0	1	2	0	1
11	1	0	0	1	1	0	0	1	0	0	0	0	0	0	1	2	1	1
12	0	1	1	0	0	0	0	2	0	0	0	0	0	0	1	2	1	1
13	0	1	0	1	0	0	1	0	0	1	0	0	1	2	1	1	7	0
14	0	1	0	1	0	0	1	0	0	1	0	0	1	4	1	1	9	0

tances. The energies of eight hydroxyl intramolecular interactions were found to yield an excellent fit(Table V-A) when the polar components of the fourteen alcohols were described by

$$V_m \delta_p^{\;2} + \frac{5}{2} \frac{V_m}{N} = \sum_i \varepsilon_{ip} R_i + \Delta_p \qquad (8)$$

where N = no. of hydroxyl groups per molecule,
ε_{ip} = i-th polar hydroxyl interaction energy,
Δ_p = polar constant,
and R_i = i-th RING interaction.

The important interactions were limited to methyls, methylenes, and hydroxyls. All RING sizes considered were important. The multiple correlation was 0.9999. The fractional coefficient 5/2 to V_m/N significantly improved the fit when V_m/N was used to account for the volume dependence of multiple hydroxyl functionality. Five degrees of freedom were allowed for the regression. The constant, Δ_p, is an energy term common to the polar component of all the alcohols.

B. Hydrogen Bonding

Hydrogen bonding is vaguely composed of a multitude of localized dipole-dipole interactions. More certain is the fact that the interactions occur over very short spatial distances relative to the atom to atom distance of polar interactions. The energies of seven hydroxyl interactions were sufficient to yield a good fit (Table V-B) using

$$V_m \delta_h^{\;2} + \frac{9}{2} \frac{V_m}{N} = \sum \varepsilon_{ih} R_i + \Delta_h \qquad (9)$$

Again, the coefficient of 9/2 to V_m/N significantly improved the fit. The smaller three-RING and four-RING interactions are most important. However, the five-RING and six-RING interactions are necessary. Methylene interactions are significant to an undetermined distance down the chain. The multiple correlation was 0.9987. Seven degrees of freedom were allowed for the regression.

C. Dispersion

The dispersion energy should have a source independent of the molecular dipoles i.e. only energy associated with the oscillating bond moments. Hansen derived all of the dispersion components from homomorphs via relationships at reduced temp-

TABLE V-A

Polar Components

No.	Meas. ΔE_p $(V_m\delta_p^2 + \frac{5}{2}\frac{V_m}{N})$	Calc. ΔE_p $(\Sigma\ \varepsilon_{ip} + \Delta_p^{(a)})$	$\frac{5}{2}\ \frac{V_m}{N}$	Meas. δ_p (b)	Calc. δ_p	Dev.
1	1006.9	999.3	188.0	3.3	3.28	-0.02
2	946.2	943.3	228.8	2.8	2.79	-0.01
3	800.1	800.6	272.5	2.2	2.20	0.0
4	851.3	851.5	308.0	2.1	2.10	0.0
5	646.6	636.3	318.0	1.6	1.58	-0.02
6	1649.4	1635.6	92.0	4.6	4.58	-0.02
7	1696.8	1705.7	69.8	5.4	5.41	0.01
8	2770.9	2268.4	112.4	4.9	4.90	0.0
9	2612.7	2615.1	61.1	5.9	5.90	0.0
10	883.2	888.1	192.0	3.0	3.01	0.01
11	951.3	965.1	230.0	2.8	2.83	0.03
12	959.6	963.4	232.0	2.8	2.80	0.0
13	798.0	800.6	394.3	1.6	1.60	0.0
14	803.6	800.6	479.5	1.3	1.29	-0.01

ΔE_p: std. dev. = 11.4 cal.
% error = 0.94%
F value = 4670.5
Multiple correlation = 0.9999

δ_p: std. dev. = 0.01
% error = 0.3%

RING Interaction	R_3^2	R_4^2	R_4^3	R_5^2	R_5^3	R_5^H	R_6^2	R_6^H
Regression Constant	653.3	365.6	618.7	273.1	329.6	9.8	-144.9	171.6
t value	47.3	29.5	40.2	20.2	29.6	1.9	-15.5	13.6

(a) Δ_p = -345.0 calories

(b) Hansen's modified values

TABLE V-B

Hydrogen Bonding Components

No.	Meas. ΔE_h $(V_m \delta_h^2 + \frac{9}{2}\frac{V_m}{N})$	Calc. ΔE_h $(\Sigma\ \varepsilon_{ih} + \Delta_p^{(a)})$	$\frac{9}{2}\frac{V_m}{N}$	Meas. $\delta_h^{(b)}$	Calc. δ_h	Dev.
1	5771.6	5556.0	338.4	8.5	8.33	-0.17
2	5836.8	5556.0	411.8	7.7	7.50	-0.20
3	5530.7	5556.0	490.5	6.8	6.82	0.02
4	5921.0	6120.5	554.4	6.6	6.72	0.12
5	5151.6	5209.0	572.4	6.0	6.04	0.04
6	9730.7	9557.6	165.6	11.4	11.30	-0.10
7	9125.6	9340.3	125.6	12.7	12.85	0.15
8	10113.8	10155.4	202.3	10.5	10.52	0.02
9	15099.9	15058.3	110.0	14.3	14.28	-0.02
10	5260.8	5219.2	345.6	8.0	7.97	-0.03
11	5051.7	5209.0	414.0	7.1	7.22	0.12
12	6063.6	6120.5	417.6	7.8	7.84	0.04
13	6014.6	6014.6	709.7	5.8	5.80	0.0
14	5468.2	5468.2	863.1	4.9	4.90	0.0

ΔE_h: std. dev. = 215.3 cal.
% error = 3.0%
F value = 330.6
Multiple correlation = 0.9987

δ_h: std. dev. = 0.11
% error = 1.2%

RING Interaction	R_3^2	R_4^1	R_4^2	R_4^3	R_6^H	R_7^2	R_{8+}^2
Regression Constant	2070.7	2277.9	1713.5	1723.7	581.1	1005.0	-273.2
t value	9.9	15.3	10.5	11.2	4.2	2.0	-1.8

(a) Δ_h = 1771.8 calories

(b) Hansen's modified values

TABLE V-C

Dispersion Components

No.	Meas. ΔE_d ($V_m \delta_d^2$)	Calc. ΔE_d ($\Sigma\ \varepsilon_{id} + \Delta_d$[a])	Meas. δ_d	Calc. δ_d	Dev.
1	4516.7	4448.7	7.75	7.69	-0.06
2	5581.1	5569.4	7.81	7.80	-0.01
3	6648.6	6594.2	7.81	7.78	-0.03
4	7304.5	7357.3	7.70	7.73	0.03
5	7097.9	6991.9	7.47	7.41	-0.06
6	4997.3	4895.7	8.24	8.16	-0.08
7	3797.9	3771.6	8.25	8.22	-0.03
8	5898.3	6018.4	8.10	8.18	0.08
9	5246.2	5251.1	8.46	8.46	0.00
10	4554.5	4540.2	7.70	7.69	-0.01
11	5454.7	5572.7	7.70	7.78	0.08
12	5081.7	5115.8	7.40	7.42	0.02
13	10864.0	11062.5	8.30	8.38	0.08
14	14185.5	14041.4	8.60	8.56	-0.04

ΔE_d: std. dev. = 132.1
% error = 2.0%
F value = 975.3
Multiple correlation = 0.9994

δ_d: std. dev. = 0.06
% error = 0.7%

RING Interactions	R_4^3	R_6^2	R_6^H	C^3	C^2	C^1	N
Regression Constants	457.0	-833.3	-368.7	1479.5	1489.4	677.0	792.8
t value	5.7	-4.2	-5.1	14.2	41.0	9.8	

(a) Δ_d = 792.8 calories

eratures. As others have suggested there is insufficient reason to believe that only London-type energies were being compared at those temperatures. Examination of Table V-C shows that, in addition to considering the usual Small-type summation over molecular moieties, several RING interactions were found to be significant. The general equation is

$$V_m \delta_d^2 = \Sigma \; \varepsilon_{id} \; R_i + \Delta_d \; . \qquad (10)$$

Again, the fit is good and appears to be independent of branching or multifunctionality. The multiple correlation coefficient is 0.9994. Seven degrees of freedom were allowed for the regression.

V. DISCUSSION

To grasp the meaning of Eqns. 8, 9, and 10 , the source of the data must be examined. Hansen borrowed δ_d from the cohesive energies of specific choices for homomorphs. A homomorph of a solvent containing a polar group is that saturated hydrocarbon that most nearly resembles it in size and shape at the same reduced temperature.

The resulting dispersion component should qualify as a first guess of the molecular association due to dispersion energies. Subsequently, he assumed the vectorial relationship would provide the combined sum of the squares of polar and hydrogen bonding components rewritten as(10)

$$\delta_{TOT}^2 - \delta_d^2 = \delta_p^2 + \delta_h^2 . \qquad (11)$$

Then a large number of polymer-solvent swelling observations allowed him to discriminately assign relative values to δ_p and δ_h for most common solvents. In a later paper(11), he modified the relative values of δ_p and δ_h due to the apparent validity of Bottcher's equation in the form of

$$\frac{W}{V_m} = \delta_p^2 = \frac{12108}{V_m^2} \left(\frac{\varepsilon - 1}{2\varepsilon + M_D^2} \right) (N_D^2 + 2)\mu \qquad (12)$$

where ε = dielectric constant,
N_D = index of refraction for sodium-D line,
and μ = dipole moment (Debyes).

Thus the modified SPC values used for this paper were not directly measurable nor observable. The specific values for the SPC are all subject to doubt. From the qualitative sources of origination, it is difficult to believe that pure dispersion, polar, and hydrogen bonding energies are being described by the respective components. However, Hansen's data appear to be internally consistent as revealed by the RING method using subsets of the same interaction matrix to describe all three components.

The recent literature has many references to the dimer and pseudo polymeric formations of pure alcohols and alcohols with various mixed cosolvents under the general heading of hydrogen bonding. Investigation by spectroscopic techniques and thermodynamic techniques has generated a great amount of knowledge about these molecular associations. However, specific steric substituent effects have only been used to describe parts of the complete solvent properties (i.e. Taft substituent polar constants, etc.). The RING method can reveal specific substituent effects to yield a complete description of Hansen's SPC. The differentiation of Hansen's polar, hydrogen bonding, and dispersion components follows.

At first glance the RING method seemingly accounts for only intramolecular interactions between the oxygen atom of a hydroxyl group and various hydrogen atoms. The mechanics used to obtain the interaction matrix certainly contribute to this simplistic view. Up to this point the detailed significance of the author's concept has not been clarified. Thus, the importance of each RING interaction energy for each SPC will now be shown to be of physical significance. The RING interaction energies of Tables V-A,B,C have been condensed into Table VI to help elucidate the importance of each interaction.

A. Polar

The values of the polar RING interactions are explained by the shielding effect which the various substituents have in protecting the oxygen atom from bonding. The single methylene hydrogen atom of R_3^1 has a negligable shielding effect as do R_4^1 and R_5^1. All primary alcohols have the methylene R_3^2 as the largest (653.3 cal) and most important (t= 47.3). Rotation about the carbon-oxygen bond allows the methylene hydrogens to sweep a cone of 52° at an average distance of about 2 Å from the oxygen or protect about 15% of the approach path to the oxygen. Refer to Figure 3. Reference is made to the center-to-center equilibrium distances. The methylene hydrogens of R_4^2 increase this cone of shielding to 134° but are located an increased average distance of 3 Å away from the oxygen to give a smaller value (365.6 cal) but still quite significant (t= 29.5). The methyl hydrogens of R_4^3 form a similar cone as R_4^2, however, the relatively free rotation of the methyl group, with its high rotational energy, about the carbon-carbon bond provides more 'occupied' space and is therefore a greater

TABLE VI

Interaction Energies of the Solubility Parameter Components

Group	Ring	Dispersion	Polar	Hydrogen Bonding
$-\overset{\vert}{\underset{\vert}{C}}H$	R_3^1	-	-	-
$>CH_2$	R_3^2	-	653.3	2070.7
$-\overset{\vert}{\underset{\vert}{C}}H$	R_4^1	-	-	2277.9
$>CH_2$	R_4^2	-	365.6	1713.5
$-CH_3$	R_4^3	457.0	618.7	1723.7
$-\overset{\vert}{\underset{\vert}{C}}H$	R_5^1	-	-	-
$>CH_2$	R_5^2	-	273.1	-
$-CH_3$	R_5^3	-	329.6	-
-OH	R_5^H	-	9.8	-
$>CH_2$	R_6^2	-833.3	-144.9	-
$-CH_3$	R_6^3	-	-	-
-OH	R_6^H	-368.7	171.6	581.1
$-CH_2$	R_7^2	-	-	1005.0
$-CH_2$	R_{8+}^2	-	-	-273.2
-OH	N	792.8	-	-
$-CH_3$	C^3	1479.5	-	-
$>CH_2$	C^2	1489.4	-	-
$-\overset{\vert}{\underset{\vert}{C}}H$	C^1	677.0	-	-
	Δ	-792.8	-345.0	1771.8

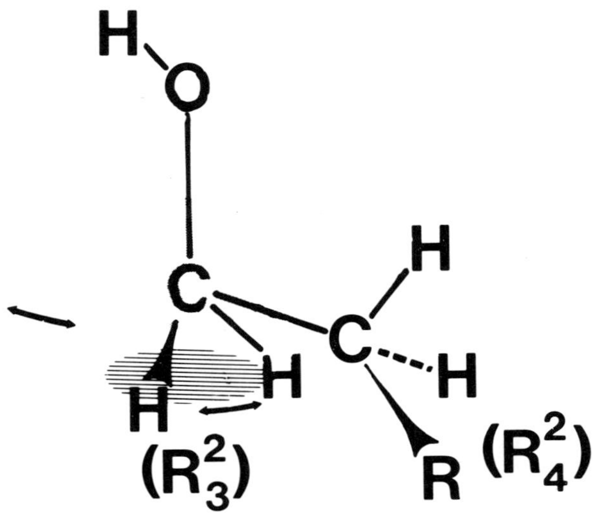

Fig. 3. The bases of the cones of protection are formed by the closest approach of the methylene hydrogens to the oxygen. The vertex is centered on the oxygen.

deterent to attack. The value (618.7 cal) is second (t= 40.2) in importance.

The methylene hydrogens of R_5^2 increase the cone to 202° with a nearest approach distance of 1.7 Å. The comparison of angle and distance with R_4^2 would imply a larger screening effect. However, another degree of rotational freedom is obtained around a carbon-carbon axis in addition to the carbon-oxygen axis. These methylene hydrogens have a larger volume of space to occupy and therefore spend less time near the oxygen atom. The decreased value (273.2 cal) retains a similar significance(t= 20.2). Again the methyl R_5^3 is of greater value (329.6 cal) and slightly more significant (t= 29.6). The mutual hydroxyl screening of R_5^H is shown to be of only minor importance and is included only to be complete. The cis configuration of the two hydroxyls would bring them within 2 Å but the trans configuration probably dominates.

The negative interaction energy (-144.9 cal) of R_6^2 is a result of intramolecular hydrogen bonding of an induced nature. An approximate pentagonal RING formation brings the methylene hydrogens to within 0.6 Å of the oxygen at the closest approach. Although this could be an effective shield, the approach causes dipole-induced dipole intramolecular hydrogen bonding and thus yields a negative sign. The interaction is weak due to the chain extension arising from an additional rotational degree of freedom about a carbon-carbon axis. Alternately, the substituent can be considered to occupy a larger volume of space as a less effective shield.

The methyl hydrogens of R_6^3 deserve a special explanation even though their effect is insignificant. The previously discussed methyl vs. methylene energies compared as $R_4^3/R_4^2 = 1.69$ and $R_5^3/R_5^2 = 1.21$. With larger RING sizes, the methyl hydrogens display a decreasingly effective shield. Now as R_6^2 hydrogens can bond with the oxygen and only have a finite shielding effect, R_6^3 rapidly falls to a negligible shield without a chance for bonding in the pure solvent case. Solvent mixtures would probably increase its significance as a negative energy. Finally, the hydroxyl interaction of R_6^H could also form intramolecular hydrogen bonding with a distance of 0.8 A. This mutual attraction could not be simultaneous due to the fact that the directional character of the modified SP^3 hybrid bonding of the oxygen allows one hydrogen bond at a time. This is reflected in the small (171.6 cal) but significant (t= 13.6) energy. Indeed, the pure solvent case leads to chain extension reducing the multiple hydroxyls to mutual screening.

The polar constant, Δ_p, is negative (-345.0 cal) and common to all of the alcohols. This corresponds to the bonding of the O-H polarized hydroxyl bond itself. This stable bond is the strongest possible.

B. Hydrogen Bonding

The accepted concept of hydrogen bonding entails a strong dipole-dipole composite attraction as depicted in Figure 4. When hydrogen is bonded to a highly electronegative atom, exposing the hydrogen nucleus, the strong positive charge of the thinly veiled proton is strongly attracted by the negative charge of an electronegative atom. This simplistic attitude arrives at the correct range of energy (5000 cal) for the alcohol hydrogen bond, but neglects specific contributions from the remainder of the molecule. Hansen was able to roughly account for his hydrogen bonding SPC by assigning a value of 4650 calories to each hydroxyl.

Careful analysis of the RING interactions leads this author to propose a detailed theory explaining Hansen's hydrogen bonding SPC. The energy values relate directly to the participation of α and β hydrogen atoms being available for mutually induced dipoles in each of two neighboring molecules. This is in addition to the dipole-dipole effect as evidenced by

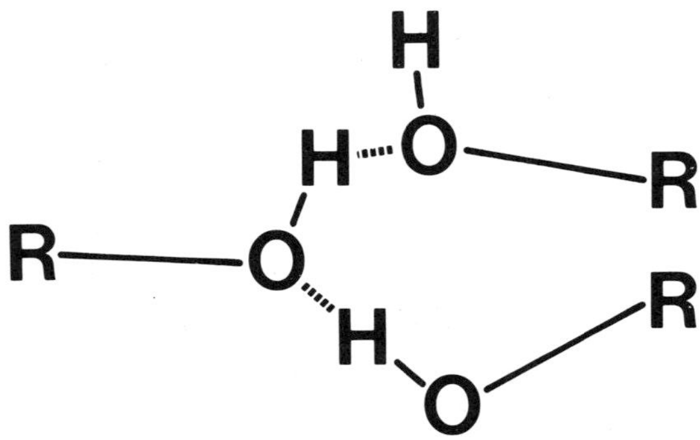

Fig. 4. Classical concept of hydrogen bonding in the general aliphatic alcohol case.

the hydrogen bonding constant, Δ_h (1771.8 cal). Figure 5 details the two mutual α hydrogen and two mutual β hydrogen possibilities for participating in an intermittent intermolecular cyclic formation. The probability of any cyclic formation is directly related to the total number of discrete 6, 7, and 8 membered formations possible.

The two α hydrogens of R_3^2 and the one β, two β, and three β hydrogens of R_4^1,R_4^2, and R_4^3, respectively, contribute 2278,1713, and 1724 calories. These values are in the same range as the dipole-dipole hydrogen bonding constant. The value for R_6^H is a fourth of the previous values (581 cal) and is due to internal hydrogen bonding between multiple hydroxyls as discussed under the polar interactions. The first intramolecular methylene hydrogens that can actually touch the oxygen atom is due to R_7^2. Here dipole-dipole hydrogen bonding results in a large RING interaction energy (1005 cal). All methylene hydrogens greater than R_7^2 are grouped together as R_{8+}^2. They can all interrupt the α, β hydrogen participation thus yielding a negative value (-273 cal).

C. Dispersion

The values of the dispersion RING interactions correlate well with the aspects of intramolecular disorder. Essentially, the longer the chain, the more disorder. The C^2 carbons containing methylene hydrogens are the major contributors followed

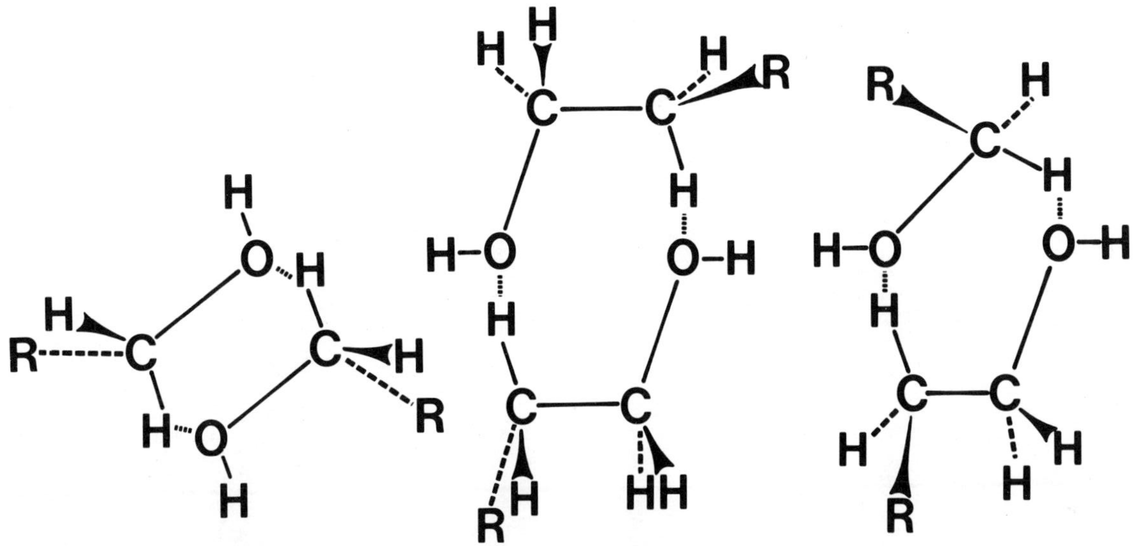

$\alpha(R^2_3)$ $\beta(R^2_4)$ α,β

Fig. 5. Additional cyclical stabilization imparted by the dipole induced-dipole α, β-hydrogen participation.

by C^3 and C^1. The only other positive value, due to the freely rotating methyl group, arises from $R_4{}^3$ which has been previously discussed. The negative contributions to disorder are the two intramolecular hydrogen bonding interactions of $R_6{}^2$ (-833.3 cal) and $R_6{}^H$ (-368.7 cal). The dispersion constant, Δ_d is negative (-792.8 cal) and represents the intramolecular ordering into hydrophilic and hydrophobic moieties by the mere presence of the hydroxyl group.

D. Ideal vs. Associated Liquids

The vibrational and rotational contributions to the internal pressure and due consideration to the degrees of freedom has allowed Bagley(13) to accurately calculate the internal pressure of several suitable solvents and suggests the following argument. If the ideal gas PVT relationship

$$PV/n = RT \tag{13}$$

is construed to represent the ideal portion of a liquid, where $V = V_m$ = molar volume, n = N = no. of hydroxyl groups contained in that molar volume, and conveniently allow P = 1, then

$$V_m/N = RT \ . \tag{14}$$

This argument, admittedly lacking sophistication, does allow a consistent interpretation of V_m/N as an energy term in Eqns. 8, 9, and 10. Furthermore, this term can account for the ideal (non-associative) portion of the polar interaction energies. For the present, the coefficient of 5/2 in Eqn. 8 is related to the fact that an ideal gas possesses 3/2 degrees of freedom or contributes 5/2 R to its heat capacity, C_p. If this reasoning has any significance then the 9/2 coefficient in Eqn. 9 should follow a similar line of reasoning.

Indeed, an ideal diatomic gas possesses 7 degrees of freedom and contributes 9/2 R to its heat capacity. The short distance "pair" interaction, in the hydrogen bonding case which is composed of " O•••H ", could be interpreted as the ideal (non-associative) portion of hydrogen bonding interaction energies.

Empirically, the dispersion relationship does not require a V_m/N ideal term. This is in agreement with the supposition that only non-associative interactions are being described.

There is presently insufficient published data on cyclic and aromatic alcohols to attempt a reasonable assignment of RING interactions. These systems are certainly susceptable to the RING method. Extension into ether-alcohols, ethers, and esters has been initiated. Table VII illustrates the calculation of the SPC of other multifunctional alcohols not yet reported in the literature. Hence, the detailed use of Table VI is illustrated.

VI. CONCLUSIONS

Utilizing the suggestions of Rheinhard and Lin(14) of partial molar volumes has allowed reasonable SPC values to be calculated for polyvinyl alcohol, but they haven't been confirmed experimentally due to the intractable nature of this polymer type in solvent studies. When the RING interactions of other functional groups are available, many oligomers and polymer systems will be available for confirmation.

Specific hydroxyl interactions in a set of fourteen varied aliphatic alcohols were used to lend support to Hansen's three dimensional solubility parameter theory and offer new insights into the molecular dynamics of polymer-solvent interactions. For the first time, a consistent set of energy terms for the calculation of solubility parameter components via structural considerations has proven feasible.

The RING method is dependent upon and limited by a self consistent set of data. Any well characterized, empirical, solubility data should be susceptible to this technique. A correct classical structure of an alcohol can now be used as the only outside information necessary at arriving at confident values for the three SPC.

The polar RING interactions specifically relate to the effectiveness of a substituent in shielding the oxygens from intermolecular approach of a foreign nucleus. Internal bonding, as it effects the potential shield, is also described. The hydrogen bonding RING interactions detail the α, β-hydrogen participation in 6, 7, and 8 membered intermolecular RING formations in addition to the more common dipole-dipole bonding. The dispersion RING interactions measures the tendency for intramolecular disorder. The percent errors in predicting the individual components (dispersion, polar, and hydrogen bonding) are 0.6%, 0.3% and 1.2%, respectively.

The analysis presented reveals that Hansen's assignments of SPC, although useful in a practical nature, do not represent pure, independent components. The homomorph theory does not give an independent value for dispersion forces. However, with the information presented here, the author believes that orthogonal SPC could be established. In summary, the RING method can account for the solubility parameter components of alcohols. Statistical analyses have shown the predicted accuracy to be independent of branching and multifunctionality. Thus, the adaptation to polymers appears feasible. The final

TABLE VII

SPC Predictions for Butanediols

RING Interactions	R_3^1	R_3^2	R_1^1	R_4^2	R_4^3	R_5^1	R_5^2	R_5^3	R_5^H	R_6^2	R_6^3	R_6^H	N	C^3	C^2	C^1	Density[(a)] (20°C)	V_m
1,2-butanediol	1	1	1	2			1	1	2		1		2	1	2	1	1.0024	89.90
1,3-butanediol*	1	1		2	1	1	1				1	2	2	1	2	1	1.0053	89.64
2,3-butanediol	2		2		2			2	2				2	2	0	2	1.0033	89.82

(a) Racemic (dl) densities: Handbook of Chemistry & Physics - 1975 Edition

	Calc. δ_d	Calc. δ_p	Calc. δ_h	Resultant δ_{TOT}
1,2-butanediol	7.86 ± 0.06	4.14 ± 0.01	10.20 ± 0.11	13.53
1,3-butanediol*	8.19 ± 0.06	4.90 ± 0.01	10.54 ± 0.11	14.22
2,3-butanediol	8.19 ± 0.06	4.02 ± 0.01	10.32 ± 0.11	13.77

Example Calculation for 1,3-butanediol

$$\delta_d = [(\Sigma\ \varepsilon_{id} - 792.8)/89.64]^{1/2} = 8.19$$

where $\Sigma\ \varepsilon_{id} = [(457.0 - 368.7 + 1479.5 + 2\ (1489.4) + 677.0 + 2\ (792.8)] = 6809.2$

$$\delta_p = [(\Sigma\ \varepsilon_{ip} - 345.0 - 112.0)/89.64|^{1/2} = 4.90$$

where $\Sigma\ \varepsilon_{ip} = [653.3 + 2\ (365.6) + 618.7 + 273.1 + 2\ (171.6)] = 2619.5$

$$\delta_h = [(\Sigma\ \varepsilon_{ih} + 1771.8 - 201.7)/89.64)]^{1/2} = 10.54$$

where $\Sigma\ \varepsilon_{ih} = [2070.7 + 2\ (1713.5) + 1723.7 + 2\ (581.1)] = 8383.6$

*Reported as #8 in this paper.

relationships are

$$V_m \delta_d^2 = \Sigma\, \varepsilon_{id}\, R_i + \Delta_d$$

$$V_m \delta_p^2 + \frac{5}{2}\frac{V_m}{N} = \Sigma\, \varepsilon_{ip}\, R_i + \Delta_p$$

$$V_m \delta_h^2 + \frac{9}{2}\frac{V_m}{N} = \Sigma\, \varepsilon_{ih}\, R_i + \Delta_h.$$

VII. ACKNOWLEDGEMENTS

The author is indebted to several previous colleagues at the Glidden-Durkee research center in Cleveland, Ohio. Mr. J. Erickson has freely given his encouragement, knowledge of solubility parameters, and advice on statistical methods. Dr. T. Provder has contributed editorial comments on the manuscript.

VIII. REFERENCES

1. A.F.M. Barton, Chem. Rev. 75, No. 6, 731 (1975).
2. C.M. Hansen and A. Beerbower, Encyclopedia of Chemical Technology, Supplement Vol., John Wiley & Sons, Inc., 1971.
3. J. Hildebrand and R. Scott, "The Solubility of Non-Electrolytes," 3rd Edition, Reinhold Publishing Corp., N.Y., 1949.
4. P.A. Small, J. Appl. Chem. 3, 71 (1953).
5. N. Nygard, Unpublished paper presented at the Western Coatings Societies' Symposium, L.A., Feb. 1964.
6. R.F. Blanks and J.M. Prausnitz, Ind. Eng. Chem., Fundam. 3, 1 (1964).
7. J.L. Gardon, J. Paint Tech. 38, No. 492, 43 (1966).
8. K. Hoy, J. Paint Tech. 42, No. 541, 76 (1970).
9. K.Hoy, "Tables of Solubility Parameters", Union Carbide Corp., Chemicals and Plastics, R&D Dept., 1969.
10. C.M. Hansen, J. Paint Tech. 39, No. 505, 104 (1967).
11. C.M. Hansen and K. Skaarup, J. Paint Tech. 39, No. 511(1967)
12. Koenhen and Smolders, J. Appl. P. Sci. 19, 1163 (1975).
13. E.B. Bagley, T.P. Nelson, and J.M. Scigliano, J. Paint Tech. 43, No. 555, 35 (1971).

STRUCTURAL STUDIES OF POLYMERS BY ENTHALPIES OBTAINED FROM HEATS OF SOLUTION

F. E. Filisko, H. T. Kau, J. DeRudder, J. Golba
University of Michigan

I. INTRODUCTION

One of the most basic parameters which can be used to characterize the state of a solid is the enthalpy. In it must be contained information about any structure which exists in the solid. It is, however, only possible to obtain relative enthalpies, i.e., enthalpy differences with respect to some reference state. For these differences to be meaningful, the reference state must be reproducible and a true thermodynamic state. Thus, if we are interested in obtaining enthalpies characteristic of the solid state of a polymer, it would be reasonable to choose the amorphous liquid state (ALS) of the polymer as the reference state since it is "structureless". It is impossible to transform a solid to a liquid below T_m or T_g directly. However, this transformation is performed routinely in the solution process and heats of solution are then a direct measure of the enthalpy content of the polymer with the dilute solution as the reference state. The dilute solution, for most systems, is an excellent reference state since it is reproducible and a true thermodynamic state, i.e. the thermodynamic state of the solution at a given concentration and temperature is identical independent of the enthalpic state of the solute. If desired, the enthalpic state of the ALS can be deduced either directly from heat of dilution vs. concentration measurements (1) or indirectly from osmotic pressure measurements using a consistent theory of solutions (2). However, if only differences in enthalpy content of polymers is desired, these can be measured directly by the difference in 2 heat of solution measurements without regard for the "heat of mixing".

Systems particularly attractive for these measurements are athermal systems or those with very low heats of mixing. In these cases, the enthalpy content of the polymer, with respect to its amorphous liquid state and heats of solution, are identical. Such systems are polystyrene in toluene or ethylbenzene, polymethylmethacrylate in orthodichlorobenzene, and polyvinyl chloride in acetophenone. A further important feature of these systems is that heats of mixing for these polymers in other solvents may be determined by comparing heats of solution of identical polymer samples in the other solvent and in an athermal solvent.

There are many well established and widely used experimental techniques used to study characteristics of molecules or groups of molecules in solids or in solution. What, then, are the unique features of this method and what is the justification for its usage.

1. Enthalpy is a basic thermodynamic parameter and its behavior is well understood in the vicinity of various transitions.
2. For athermal systems, the measurements of enthalpy are direct requiring no corrections or additions to the measured values, thereby eliminating uncertainties which may arise when modifying data with questionable theories. For non-athermal systems, the measurements do require a correction for heat of mixing if absolute enthalpy is desired. However, these can be directly measured by heats of dilution. Usually, differences in enthalpy are all that is needed to characterize differences in structure and these, of course, are absolute enthalpy differences.
3. Heat of solution measurements are very sensitive and differences of a few tenths of calories per gram which are easily detectable could correspond, for instance, to fractions of percent crystallinity. Further, these measurements may be quantitative if a heat of fusion for the 100% crystalline material is known. In general, however, care must be taken in using heats of fusion for this purpose since the order or crystallinity seen by this method may not be the same as that observed in x-ray for instance.
4. As with x-ray, this method allows measurement of order or crystallinity at room temperature to eliminate variations which may occur upon heating. In addition, it also allows characterization of glassy materials at temperatures well below where annealing effects may produce anamolies.
5. The measurements are isothermal which means they are true enthalpy measurements, in contrast to scanning measurements which detect a behavior which is a complex function of a number of variables (3) including kinetics of a transition and scanning rate of the instrument. Further, kinetics of annealing at any temperature can be followed by the direct measurement of enthalpy change with time of annealing.

In this article we will examine unique features of this method of characterization, how the enthalpy content of a polymer may be analyzed, and characteristic behavior of the enthalpy content observed for polymer systems.

II. EXPERIMENTAL

The heats of solution are obtained by means of a Tian-Calvet type differential isothermal microcalorimeter. Sampling techniques are the most difficult part of these measurements but are described elsewhere and will not be recounted here (4,5). One point will be mentioned, however, which concerns the slow solubility of polymers. This problem is easily overcome by using small amounts of sample and in the form of very thin discs which is possible because of the extreme sensitivity of the instrumentation. Typical measurements by us involve samples of about 10 mgs in the form of thin discs 3/8 inch in diameter and about 0.1 mm thick. These are obtained either by using thin films directly or by compacting powders into this form. In most cases we use the latter method since we prefer to purify our samples by reprecipitation from which we obtain a powder, or to use non-compounded resins.

III. THEORETICAL

This analysis appears elsewhere (6) but will be partially repeated here to minimize discontinuity and to emphasize the directness of these measurements.

The integral heat of solution ΔH_s is the total heat measured calorimetrically when two distinct phases (i.e., solvent and solute) are brought together and form a single phase (i.e., solution). In our case, the solvent and solution are liquids and in different measurements the solutions may differ by only the resulting concentration. But, at the same concentration and temperature, the enthalpy of the solution is always the same. The solute, on the other hand, may be in an infinite number of enthalpic states due to varying amounts of crystallinity or severity of thermal and mechanical treatment. Thus, differences in heats of solution, if all conditions of solution are identical, are due to differences in enthalpy content of the solid and are solvent independent. The solvent dependence can be isolated from the integral heat of solution as the heat of mixing which is the enthalpy difference between the solution and the sum of the enthalpies of the solvent and "amorphous liquid state" (ALS) of the polymer. The "ALS" of the polymer, of course, cannot be obtained except above T_m. The heat of mixing, however, can be determined by osmotic pressure measurements or directly by heats of dilution. Thus, we can write,

$$H(solution) - [H(solvent) + H(solute)] =$$

$$\{H(solution) - [H(solvent) + H(ALS)]\} - [H(solute) - H(ALS)]$$

where, $H(ALS)$ = enthalpy of "amorphous liquid state" of the solute

$$H(solution) - [H(solvent) + H(solute)] = \Delta H_s = \text{heat of}$$

solution which is measured calorimetry.

H(solution) - [*H(solvent)* + *H(ALS)*] = ΔH_m = heat of mixing which contains any solvent-solute interactions

H(solute) - *H(ALS)* = ΔH_r = residual enthalpy or difference in enthalpy between the solid and the equilibrium "ALS" of the polymer, i.e., the enthalpy content of the solid polymer.

That is,

$$\Delta H_s = \Delta H_m - \Delta H_r$$

where, ΔH_r contains all information about the structure and morphology of the solid polymer.

Figure 1 can be used to graphically illustrate the significance of the "enthalpy content" of a polymer, even though a real plot of enthalpy versus temperature would not be linear above and below Tg (since this would indicate constant heat capacities above and below Tg).

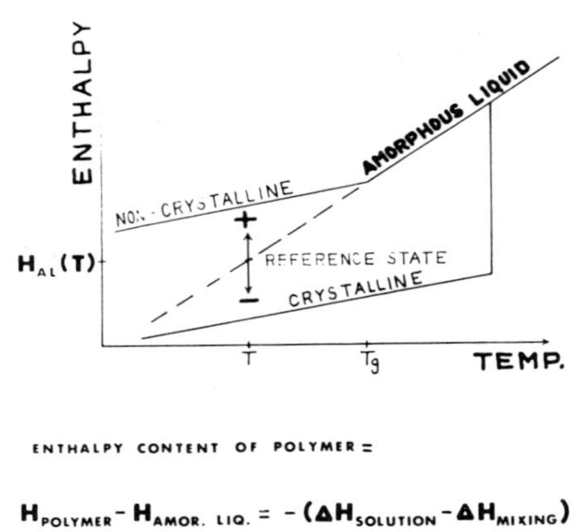

Fig. 1. Pseudo-representative plot of enthalpy versus temperature for a polymer in its crystalline and noncrystalline state.

The reference state, i.e. the ALS is illustrated by the dashed line which is the extrapolation of the amorphous liquid curve from above Tg. At any temperature, T, a positive enthalpy content means the material is in a higher enthalpic state than the ALS at that temperature. Conversely, a negative enthalpy content means the material is in a lower enthalpic state than its ALS at T. The two extremes in structure are illustrated by the lines labeled non-crystalline and crystalline. If a material is completely noncrystalline, its enthalpy content will be positive and decrease linearly with temperature to zero at and above Tg. If the material is completely crystalline, the enthalpy content will be negative and its absolute value will increase with temperature to Tm above which it will abruptly go to zero. This, of course, assumes an ideal case where no annealing affects occur. On the other hand, if a material is semicrystalline, below Tg the enthalpy content will contain a positive contribution from the glassy portion and a negative contribution from the crystalline regions. However, since the positive glassy portion is decreasing and the absolute value of the negative crystalline portion is increasing with temperature, at some point the enthalpy content must become negative, and remain so up to the Tm. In fact, above Tg, the only contribution to the enthalpy content will be from ordered or crystalline regions, assuming no transitions exist between Tg and Tm. Thus, measurements of enthalpy content versus temperature may be used and are used to characterize a material and facilitate interpretation of the data.

IV. DISCUSSION

Typically, enthalpy content versus temperature for a glassy material decreases linearly with increasing temperature below Tg and extrapolates to zero in the vicinity of Tg (5,7) as illustrated in Figure 2. To obtain information about structure from this data, we must consider various possible contributions to the enthalpy content. Contributions which make the enthalpy content of a material higher than its ALS are macroscopic inhomogeneities, nonequilibrium conformations when the material is at a temperature such that it is in a supercooled state, excess free volume or expanded coils, heat capacity difference between the solid and its ALS. Negative contributions to the enthalpy are order or crystallinity, and nonequilibrium conformations when the material is at a temperature such that it is in a superheated state.

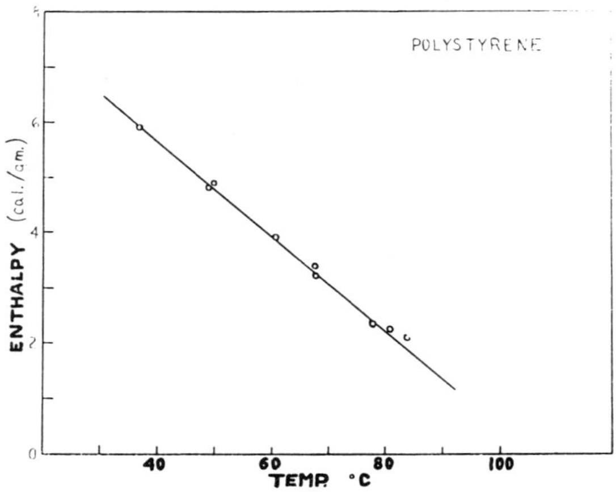

Fig. 2. Enthalpy content of atactic polystyrene versus temperature.

A. Macroscopic Inhomogeneities

These refer to frozen in strains and stresses as a result of large scale density fluctuations within a sample. This contribution is eliminated in our measurements by using samples a few mils thick and by appropriate thermal treatment.

B. Nonequilibrium Conformations

As discussed in detail by Kovacs (8), Illers (9), Petrie (10) and others, when a polymer is cooled from above to below Tg, the conformational state of the chains become frozen and retain a value corresponding to approximately Tg. With annealing time at temperatures not too far below Tg, the chains can attain lower conformational energy states which are the equilibrium conformational states at some temperature $T_2^e < Tg$ (10). At this temperature, presumably the enthalpy content of the material is zero. Thus, at any temperature $T < T_2^e$ the material is in a nonequilibrium conformational state of higher enthalpy than the ALS, i.e. it is supercooled and its nonequilibrium conformational enthalpy content is positive. For polymers in general, because of the sluggishness of the chains at temperatures well below Tg, the material will never anneal to its equilibrium conformational state. On the other hand, at temperatures slightly above T_2^e, the molecules may still be sufficiently sluggish so that initially the material is in a nonequilibrium conformational state of lower enthalpy than the ALS, i.e. it is superheated and its nonequilibrium conformational enthalpy content is negative. Because these temperatures are generally near Tg, the molecules will have sufficient mobility so that annealing can occur and the material will ultimately attain its equilibrium conformational state and have zero enthalpy content at that temperature. According to this analysis there are two important points worth recapping. First, a material at a temperature $T < T_2^e$ has a positive conformational enthalpy content and must anneal toward zero, while a material at $T > T_2^e$ initially has a negative conformational enthalpy content and must, upon annealing, increase its enthalpy and approach zero. Second, at $T = T_2^e$ the conformational enthalpy content is zero and no annealing can take place. Thus, the nonequilibrium conformational contribution to the enthalpy may be eliminated by performing measurements at $T = T_2^e$.

C. Excess Volume

Excess volumes, holes, expanded coils, etc. all must contribute positively to the enthalpy. However, measurements performed by us and others (11,12) on samples cooled from above Tg under various pressures show no significant differences in the measured enthalpies even though large differences in specific volume were observed. The measured enthalpies, in fact, give the same value as for samples quenched from 270°C at atmospheric pressure. This strongly indicates that there is apparently no relationship between specific volume and enthalpy.

D. Order

The presence of ordered regions or crystallinity in glassy polymers is an area of great debate mainly because it cannot be shown directly. If it did exist, it would provide a negative contribution to the enthalpy content. Table I illustrates the effect of orientation on enthalpy content of polycarbonate and polystyrene films hot drawn to various degrees.

The total enthalpy content is positive because the materials are presumably noncrystalline and the measurements are made well below the glass transitions of the polymers. The data shows clearly, however, that orientation or ordering contributes a negative contribution to the enthalpy thus decreasing the total enthalpy content.

TABLE 1
Enthalpy Content of Polycarbonate at 35°C and Polystyrene at 38°C Versus Hot Drawing

Area Reduction (A /A)	*Enthalpy (cal/gm)*
	Polycarbonate
undrawn	+13.9
15	+13.3
80	+10.7
90	+ 9.3
	Polystyrene
undrawn	+ 5.14
20	+ 4.51
40	+ 3.81

E. Heat Capacity Contribution

This contribution is simply due to the heat capacity difference between the glass and its ALS below Tg. It can be easily calculated (13) if heat capacity data exist and subtracted from the measured enthalpies. Attempts by us to perform these calculations using published heat capacity data have led to results which are difficult to explain. Possibly, part of the difficulty arises from the Cp data which, to be coincident with our measurements, should be measured at times consistent with out measurements. In other words, it is a characteristic of our instrumentation that a sample be held at the measurement temperature for about 1.5 hours before a measurement is made.

In the foregoing analysis we have attempted to show that despite many possible contributions to the enthalpy, only order or crystallinity in small amounts and conformational nonequilibrium contributions cannot be determined or calculated a-priori. However, we have also shown that, due to the nature of these contributions, it is possible to separate them by making measurements above Tg or At T_2^e, since in both instances the nonequilibrium conformational enthalpy content is zero with respect to the ALS.

V. RESULTS

The foregoing discussion will now be applied to analysis of enthalpy content data on atactic poly (methylmethacrylate). This data was obtained using orthodichlorobenzene as a solvent. Figure 3 shows a plot of enthalpy versus temperature for the PMMA reprecipitated from solution and dried under vacuum at 100°C. The approximation was made that this system is athermal, which is nearly the case (7,14).

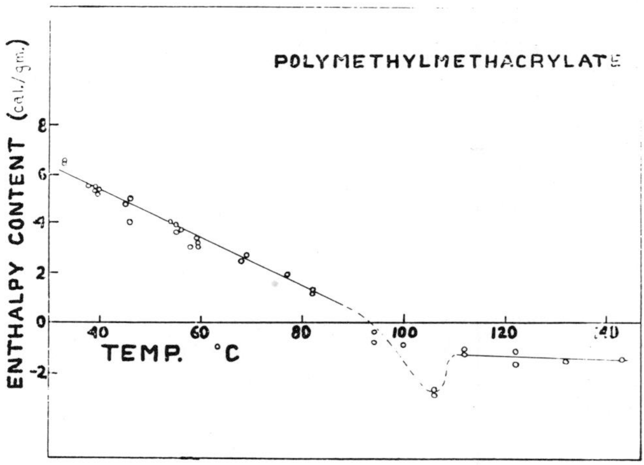

Fig. 3. Enthalpy content of annealed atactic poly (methylmethacrylate) versus temperature.

Below 90°C the enthalpy content is positive since the material is noncrystalline and below its Tg and decreases linearly to zero at about 95°C which is about 15°C below the Tg as measured by a Differential Scanning Calorimeter. This linear decrease is consistent with all other systems studied (5,7). Because no annealing effects are observed below 80°C, the slope of this line must be related to the heat capacity of the solid. Figure 4 illustrates the calculation of enthalpy content based upon the heat capacity difference between the solid and its ALS. Analysis of this calculation indicates that a linear decrease of enthalpy with temperature can be explained in a first approximation if the Cp vs. T plot for the solid and the extrapolated plot for the liquid are parallel. However, published Cp data in the majority of cases indicate that this is not the case (15).

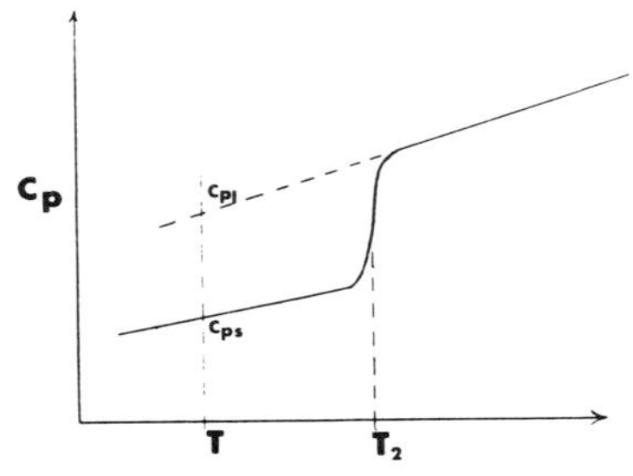

$$h_l - h_s = -\int_T^{T_2^e} (c_{pl} - c_{ps})\, dT \qquad T < T_2^e$$

$$= -\int_T^{T_2^e} \Delta c_p(T)\, dT$$

$$\text{where} \quad \Delta c_p = c_{pl}(T) - c_{ps}(T)$$

Fig. 4. Calculation of enthalpy difference between the amorphous liquid and solid states of a polymer from heat capacity data.

Above 95°C the enthalpy data become negative and appear to deviate from the lower temperature linearity. If the behavior between 95 and 110°C is real, it cannot be explained solely on the basis of superheating since our measurements are isothermal and any nonequilibrium conformational contribution must either lie on the extrapolated line or be nearer to zero, i.e. being nonequilibrium in nature, it will anneal only toward zero which is the equilibrium state. Above 110°C the material is above its Tg and there should be no contribution to the enthalpy from the ALS. However, a negative enthalpy is observed which slowly becomes more negative with increasing temperature and suggests some ordering or crystallinity. To follow this line of reasoning, if this negative enthalpy was due to ordering, then it should be possible to quench it out and observe corresponding differences in enthalpy. Figure 5 illustrates enthalpy data on samples of PMMA quenched from 200°C superimposed on the data for the annealed sample. Here it is quite clear that below 55°C, the quenched material is in a higher enthalpic state than the as-reprecipitated. However, above 55°C data for both samples superimpose. The behavior at 55°C for the quenched material may be explained by enthalpy relaxation but the narrow temperature interval over which it occurs, as well as the large temperature difference below Tg, make this unlikely. Further, the negative enthalpy above

110°C and the difference in enthalpies of the two samples below 55°C are very close, suggesting that the same mechanism may be responsible for both these enthalpies. We feel that at 55°C a crystallization is going on in the quenched sample and the order or crystals thus formed is responsible for the negative enthalpy above 110°C.

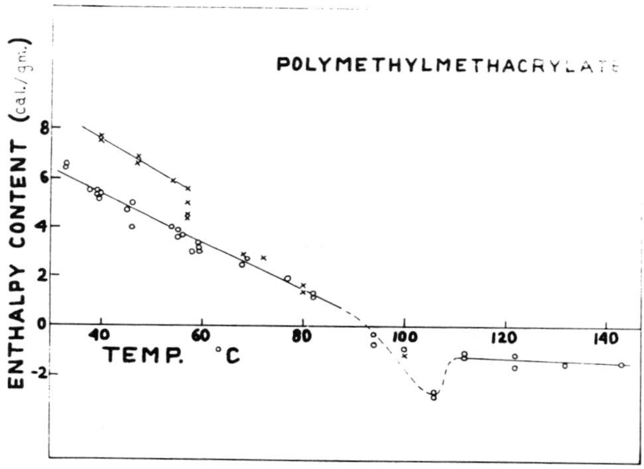

Fig. 5. Enthalpy content of quenched (x) and annealed (O) poly (methylmethacrylate).

A possible explanation of the source of this order is based upon the presence of isotactic stereochemical impurities in the seemingly atactic PMMA as reported in many articles (16, 17). Isotactic PMMA is crystallizable and its Tg is in the vicinity of 55°C. However, negligible diffusion of isotactic segments can occur at 55°C since the material is rigid. Therefore, if we are observing crystallization at 55°C, the isotactic segments must already be associated. This is possible in our samples because of the method of preparation (i.e. reprecipitation from solution) as well as the fact that isotactic and atactic PMMA are imcompatible (18).

VI. CONCLUSION

In this article we have attempted to illustrate a unique method by which mechanisms of solution and solubility properties of polymers may be used to characterize and image structures present in the solid polymers.

VII. REFERENCES

1. Maron, S. H., and Filisko, F. E., J. Macromole. Sci. Phys. 6, 413 (1972).
2. Maron, S. H., and Daniels, C. A., J. Macromole. Sci. Phys. 2, 769 (1968).
3. Garn, P. D., and Menis, O., Polym. Preprints 17 (2), 151 (1976).
4. Maron, S. H., and Filisko, F. E., J. Macromole. Sci. Phys. 6, 57 (1972).
5. Filisko, F. E., J. Macromole. Sci. Phys. 10, 371 (1974).
6. Filisko, F. E., Polym Preprints 17 (2), 419 (1976).
7. Filisko, F. E., and Raghava, R., J. Appl. Phys. 45 (10), 4155 (1974): 4151 (1974).
8. Kovacs, A. J., Fortschr. Hochpolym.-Forsch. 3, 394 (1963); J. Polym. Sci. 30, 131 (1958).
9. Illers, K. H., Makromole. Chem. 127, 1 (1969).
10. Petrie, S. E. B., J. Polym. Sci. A-2 (10), 1255 (1972).
11. Price, C., Williams, R. C., and Ayerst, R. C., in "Amorphous Materials", (R. W. Douglas, and B. Ellis, Eds.), p. 117. John Wiley and Sons, New York, 1972.
12. Allen, G., Ayerst, R., Cleveland, J., Gee, G., and Price, C., J. Polym. Sci. C28, 127 (1968).
13. Swalin, R. A., "Thermodynamics of Solids", 2nd edition, p. 28, John Wiley and Sons, New York, 1972.
14. Daoust, H., and Hade, A., Polymer 9, 47 (1968).
15. Wunderlich, B., and Baur, H., Fortschr. Hochpolym.-Forsch. 7, 151 (1970).
16. Bovey, F. A., and Tiers, G. V. D., Fortschr. Hochpolym.-Forsch. 3, 139 (1963).
17. Bovey, F. A., J. Polym. Sci. 46, 59 (1960).
18. Williamson, J. R., Proc. of 4th Annual Plastics Conference at Eastern Michigan University, (Carter, G. F., Ed.).

SOLUBILITY AND ASSOCIATION OF SOME HETEROCYCLIC POLYMERS

C.-P. Wong and G. C. Berry

Carnegie-Mellon University

I. INTRODUCTION

Over the past several years studies in our laboratory and elsewhere on the polymers BBB and BBL prepared, (1,2) respectively, by the polymerization of 3,3'diaminobenzidine or 1,2,4,5-tetraaminobenzene with 1,4,5,8-naphthalene tetracarboxylic acid (NTC) have included a wide range of topics, including the kinetics of polymerization, (3) the physical chemistry and rheology of dilute and concentrated solutions (4-8) and the viscoelastic properties of films and fibers. (9-12) We will review these results, together with some recent studies from our laboratory. As may be seen from the structures

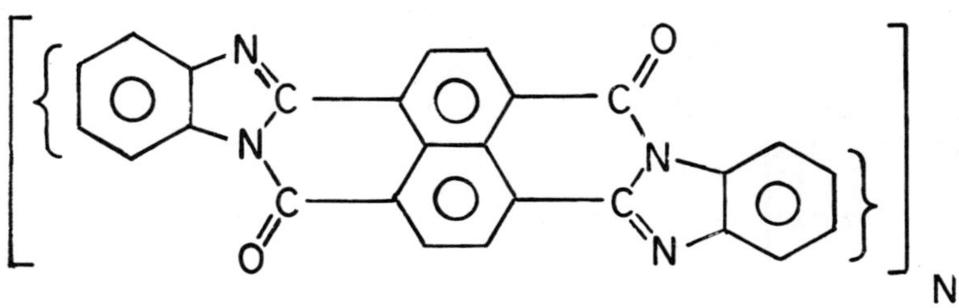

BBB

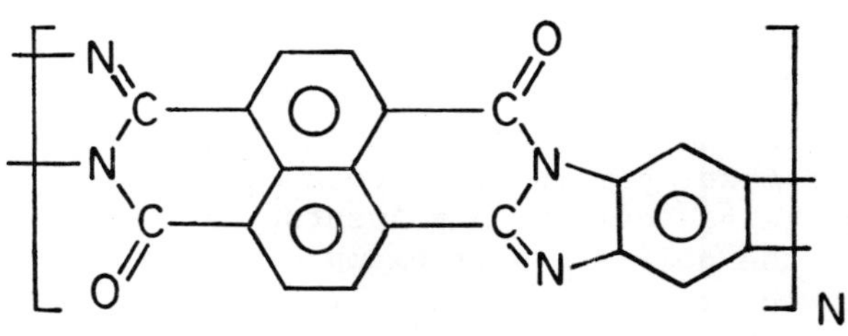

BBL

the two polymers have similar chemical structure, but BBL is a rigid chain ladder polymer whereas BBB is not.

II. THE SOLID STATE

One of the most striking features of BBB and BBL is their complete lack of solubility in organic reagents and the relative ease which they can usually (see below) be dissolved in strong protic reagents. (4) In fact, even the model compound of the BBB repeating unit (prepared by reacting two moles of o-diaminobenzene with one mole of NTC) is sparsely soluble. None-the-less, measurements of the heat of sublimation of the model compound show that the 'inner heat of sublimation' is only about 1.1 kcal/mol of ring atoms, which is on the low side of the values found for a variety of polynuclear hydrocarbons. (9) Moreover, creep, dynamic mechanical measurements, and differential scanning calorimetry on noncrystalline films of BBB show that there are no temperature transitions up to temperatures of 500°C in well annealed samples. (9,10) With such samples, the creep and recovery obey the Andrade cube-root creep relation (where $\varepsilon(t)$ is the tensile strain, σ the stress, and D_A an instantaneous compliance):

$$\varepsilon(t) = \sigma D_A[1 + (t/\tau_A)^{1/3}] \tag{1}$$

The 'time-constant' τ_A for Andrade creep is typically of the order 10^6 sec for BBB and relatively independent of temperature. This behavior, together with x-ray diffraction data showing a strong reflection with a 3.4Å Bragg spacing that can be oriented by certain processing techniques indicates that the planar repeating units are stacked in a 'graphite-like' array giving BBB and BBL in the solid state the properties of densely crosslinked polymers.

III. PROPERTIES IN SOLUTION

The pervasive intermolecular association just described must be destroyed in order to achieve solubility. This can only be achieved with solvents that interact strongly with the polymer. For example, dissolution in strong protic acids is accompanied by extensive protonation of the polymer (as many as nine protons per repeating unit with BBB (7)) even though the pK_B of the polymer is only about −7. In these solvents, it is the high density of charges that disrupts the intermolecular association and makes solubility possible. Similar specific interactions exist with other solvents for BBB and BBL, including antimony trichloride.

Once dissolved, for example in strong acids, BBB and BBL can be studied by most of the usual methods including exclusion chromatography (GPC), light scattering and viscometry. (4,5,8) In dilute solution studies it is found that BBB and BBL adopt coil and rod-like conformations, respectively. Care must be taken, however to avoid association which can sometimes lead to metastable aggregates. For example, fractions of BBB prepared by GPC with solutions in 96% sulfuric acid remain in solution indefinitely. Once precipitated, however, the fractions are no longer soluble in 96% sulfuric acid (although adopting the red color of the protonated form of the polymer) but will dissolve in methane sulfonic acid (MSA). (4)

Analogous behavior is observed with BBL fractions prepared by GPC with solutions in MSA, except that in this case no known solvent will completely eliminate association in the fractions once it has occurred. For example, the weight average molecular weight M_w determined by light scattering is shown in Fig. 1 as a function of the elution volume V_e on the preparative chromatograph for fractions of BBL for two types of preparations: In one case the solutions for light scattering were taken directly from the preparative chromatograph, unfilled points. In the other case, the polymer from the chromatograph was precipitated and redissolved for the light scattering studies. It is evident that in many cases the precipitation step led to association that was not eliminated on redissolution in MSA. It may be noted that this effect would not have been detected if only the intrinsic viscosity $[\eta]$ had been correlated with V_e, see Fig. 2. The variation of the behavior among the precipitated samples can be traced to details of the precipitation procedure. The samples showing the least amount of association were rapidly precipitated, whereas the other samples were precipitated very slowly, with the pH held for several hours in a range where the polymer was coagulated but still highly protonated.

In addition to the effect of M_w, it was noted that the depolarization of the light scattering from the solutions was markedly effected by association. The light scattering from the solutions free of association was strongly depolarized, and the variation of the effect with molecular weight was consistent with a rodlike conformation

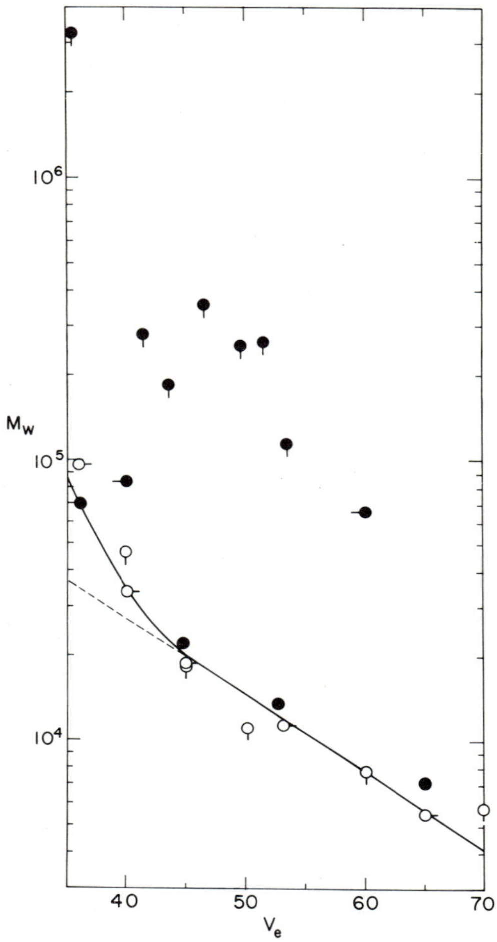

Fig. 1. Weight average molecular weight versus elution volume for solutions of BBL fractions in MSA for solutions directly from the chromatograph, unfilled circles, and solutions prepared from fractions precipitated and dried after chromatography, filled circles.

for BBL. by contrast, almost no depolarization was observed with the solutions in which association was present. This behavior suggests that the aggregates of BBL are not parallel arrays of rodlike polymer, but rather have a 'brush-

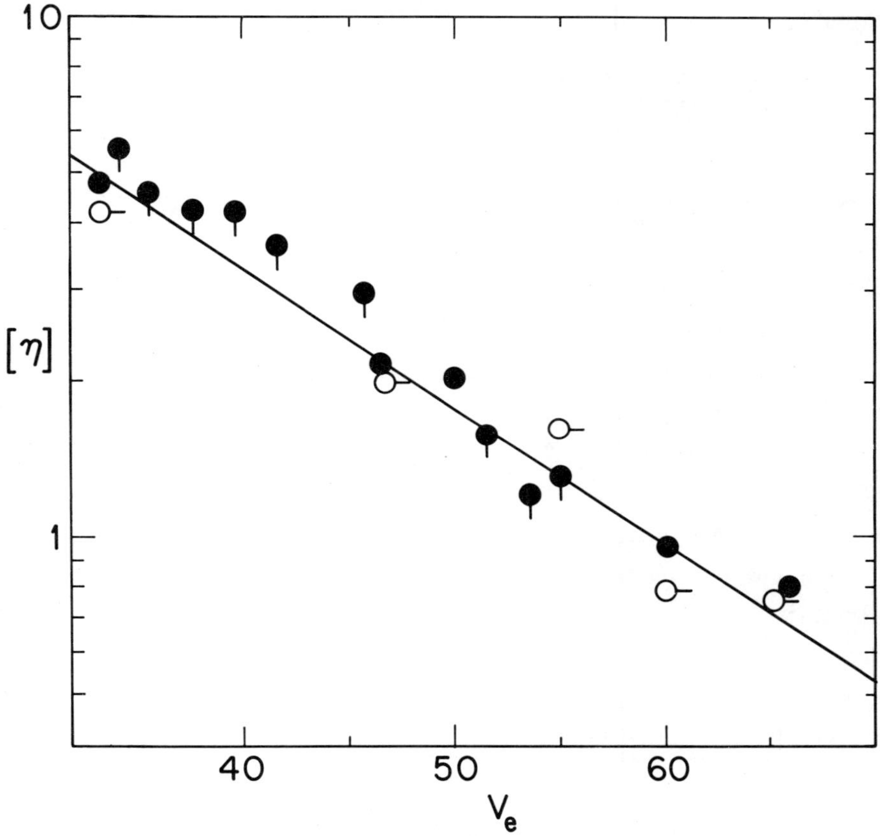

Fig. 2. Intrinsic viscosity versus elution volume for solutions of BBL fractions in MSA. The symbols are identified in the caption to Fig. 1.

heap' kind of structure.

Although association can usually be eliminated in dilute solutions of BBB, there are some circumstances when this is not so. (9) For example, heat treatment of BBB films with a certain morphology results in an incompletely soluble polymer. The insoluble fraction can be swollen to various degrees in a variety of the usual solvents for BBB, see Table I, but none of these cause further dissolution of the insoluble polymer. The nonmonotone dependence of the swelling on, say, the Hammet acidity function, shows that protonation is only one factor important in the solubility of BBB in strong acids. The variation of the molecular weight of the soluble components with the weight fraction

TABLE 1

Swelling of Insoluble BBB in Various Solvents

Solvent	Temperature (°C)	Swollen Width / Unswollen Width
Antimony trichloride	80	3.0
Chlorosulfonic acid	25	2.7
Benzene sulfonic acid	60	2.5_5
Methane sulfonic acid	60	2.4
Methane sulfonic acid	25	2.4
97% sulfuric acid	25	2.0_7
85% phosphoric acid	60	2.0_7
85% phosphoric acid	25	2.0_7
85% phosphoric acid	10	2.0_7
86% sulfuric acid	25	2.05
80% sulfuric acid	25	1.85
87% sulfuric and saturated with potassium bisulfate	25	1.35

of the gel shown in Fig. 3 is in reasonable accord with the simple theory for gel formation of a coil-like chain. This behavior has been interpreted as evidence for creation of especially stable aggregates on prolonged annealing of BBB, serving as physical crosslinks. The creep behavior of such heat treated polymers is still given by Eq. (1), but with τ_A increased by one hundred fold at room temperature (9), showing the effect of the increased intermolecular order.

The rheological properties of BBB and BBL in concentrated solutions can be studied using more-or-less standard methods. Of course, care must be taken to exclude moisture from the solution during the measurement. We have utilized capillary viscometers and a specially constructed cone and plate rheometer for rheological studies on concentrated solutions of BBB. (4,8,13) For example, the viscosity η_o

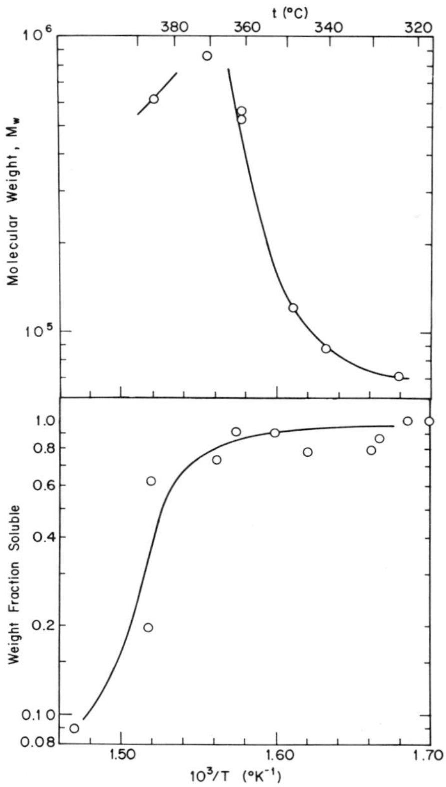

Fig. 3. The weight fraction and molecular weight of soluble fraction of BBB annealed for 15 hr. in vacuo as a function of the annealing temperature. (9)

(in the limit of small shear rate κ) is shown as a function of concentration and molecular weight in Fig. 4 for solutions of BBB in antimony trichloride. (8) Similar data have been published elsewhere for solutions of BBB in MSA. (4) In both cases, the dependence $\eta_0 \alpha (cM)^{3.4}$ typical (13) of entangled polymer chains (at temperatures far above the glass temperature) is observed for $cM > \rho M_c$ with the critical molecular weight M_c being very small (ρ is the density), of the order 200 for BBB. These very low values of M_c, which have been attributed to the effects of incipient association, (4,8) are also seen in certain cellulosic compounds. (4)

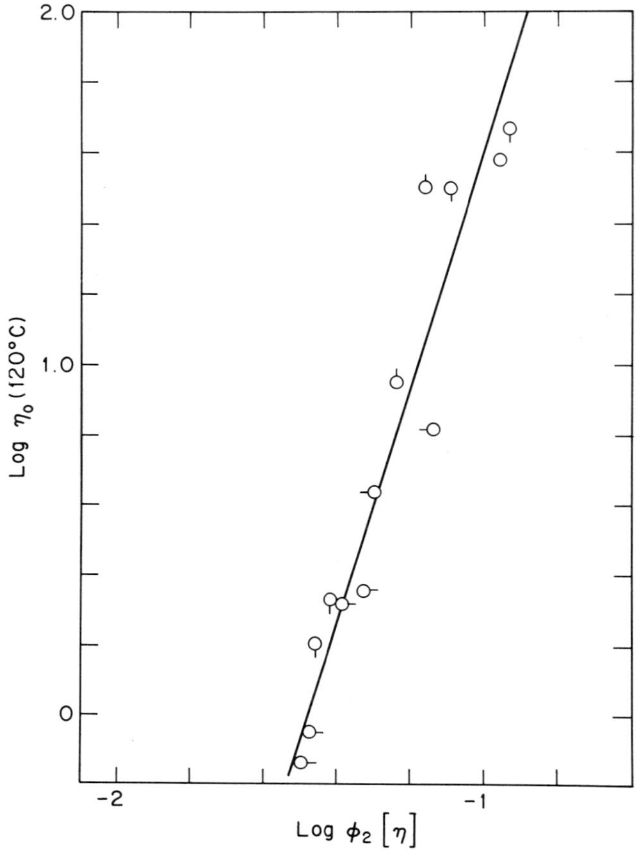

Fig. 4. Logarithmic plot of zero shear rate viscosity η_o *against the product of polymer volume fraction* ϕ_2 *and intrinsic viscosity* $[\eta]$ *for solutions of BBB in* $SbCl_3$ *at 120°C.*

Typical flow curves for the viscosity η_κ as a function of shear rate κ for solutions of BBB in MSA are shown in Fig. 5a. The reduced flow curve η_κ/η_o versus $\tau_c\kappa$ is also shown in Fig. 4b. Here the characteristic time τ_c is given by $\eta_o R_o$ where R_o is the limiting value of the steadystate recoverable compliance at small κ. The dashed line in Fig. 5b gives the reduced flow curve for a solution of polyisobutylene in cetane at a similar concentration of polymer for comparison with the BBB flow curve. The reduced flow curves are similar, as should be expected if intermolecular association is absent, see below. Consequently, as with most polymers, the viscosity of BBB solutions can

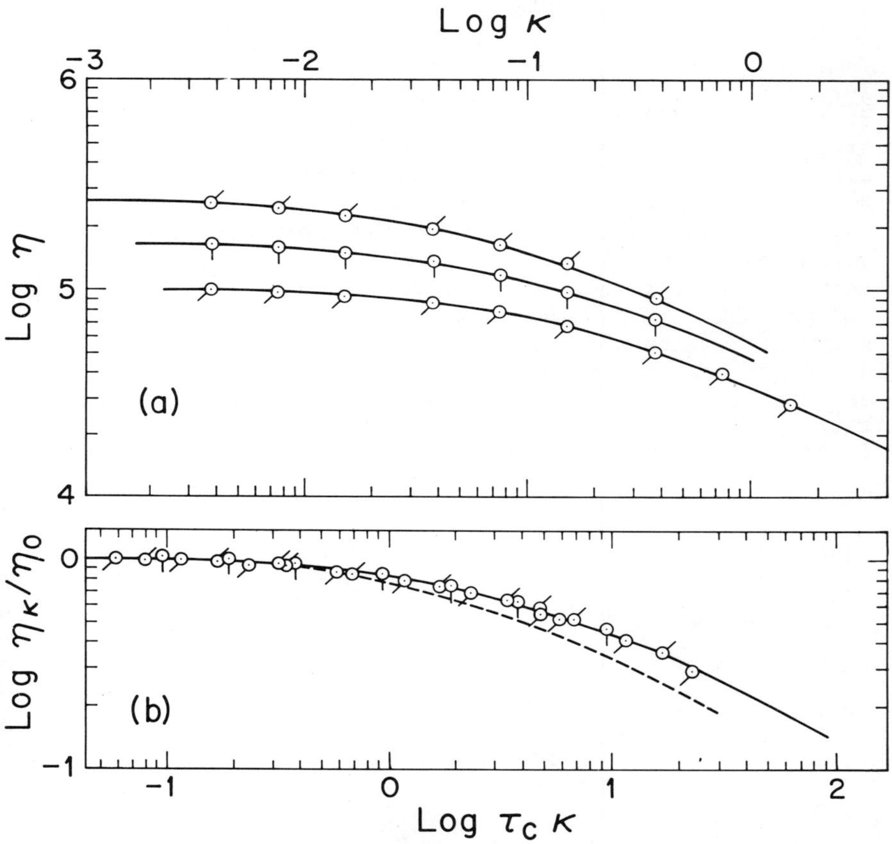

Fig. 5. a. Steady state viscosity η *plotted logarithmically against rate of shear* κ *for a solution of BBB in MSA (M_w = 6.5 x 10^4, c = 0.175 g/mℓ) at three temperatures. Key: Pip clockwise from two o'clock: 24.0°, 32.8°, and 41.2°C. b. Reduced plot* η_κ/η_o *versus* $\tau_c\kappa$*; Dashed curve, PIB/cetane.*

be represented by the relation

$$\eta_\kappa = \eta_o Q(\tau_c \kappa) \tag{2}$$

IV. ASSOCIATION IN CONCENTRATED SOLUTIONS

The effect of small amounts of moisture in causing reversible association with BBB and BBL is sometimes frustrating, but also affords an excellent opportunity to study association *per se* since association can be effected with negligible effect on other parameters (e.g., it is not necessary to change the temperature to alter the degree of association, causing other, perhaps ill-defined variations in the properties of the system). Water is very effective in causing association in solutions of BBB and BBL in MSA or sulfuric acid because it is a stronger base than the polymer, thereby causing deprotonation and consequent association of the polymer. The effects of water in the electronic absorption of a solution of BBB in MSA (0.176 g/mℓ) are shown in Fig. 6. An initially dry solution mounted in an enclosed attenuated

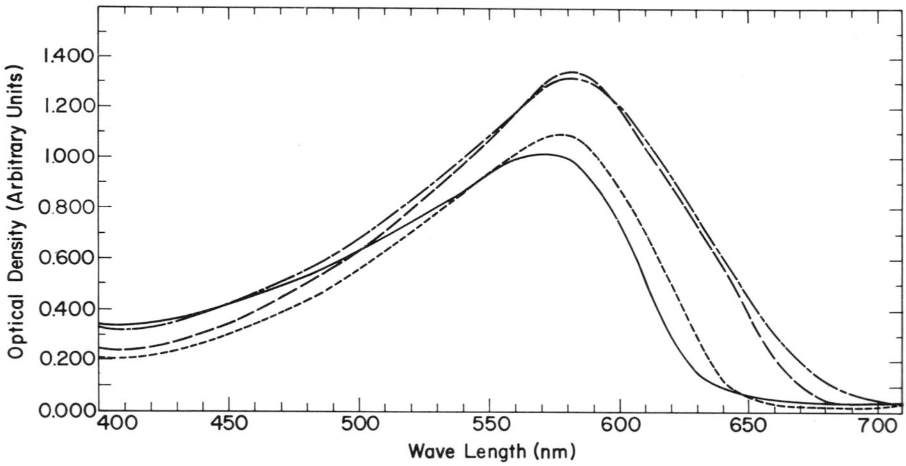

Fig. 6. ATR spectra of the BBB/MSA solution at room temperature at various aging times: ——— *zero,* ---- *10 hrs.* — — — *17.5 hrs. and* —·—·— *24.0 hrs.*

total reflectance (ATR) apparatus was allowed to absorb water slowly over a long period of time. The shifts in the spectrum can be attributed to the effects of deprotonation and association of the deprotonated solute. (15)

The effects of association on the rheological properties of concentrated solutions can be quite large. For example, the increase in η_o with aging time for a solution exposed to an atmosphere with a low partial pressure of water is shown in Fig. 7. The effect of temperature on the

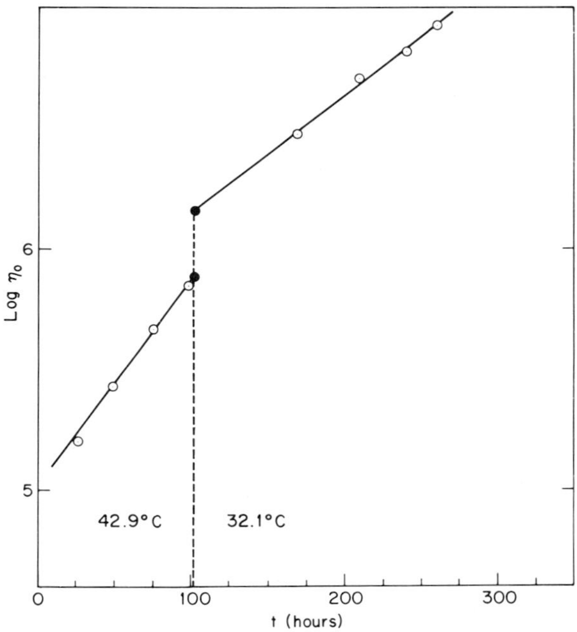

Fig. 7. Logarithm of zero shear viscosity log η_o versus aging time t at two temperatures 42.9°C and 32.1°C for the BBB/MSA solution.

rate of change of η_o can be attributed to the decreased segmental mobility with decreasing temperature. The effect of moisture is reversible, at least at low levels of water. Thus, the data in Fig. 8 show that η_o and $\tau_c\eta_o^{-1}$ both increase when a flow of dry argon over the sample is decreased, allowing the sample to adsorb moisture as it slowly leaks into the apparatus from the air, but both parameters decrease as the argon flow is re-established.

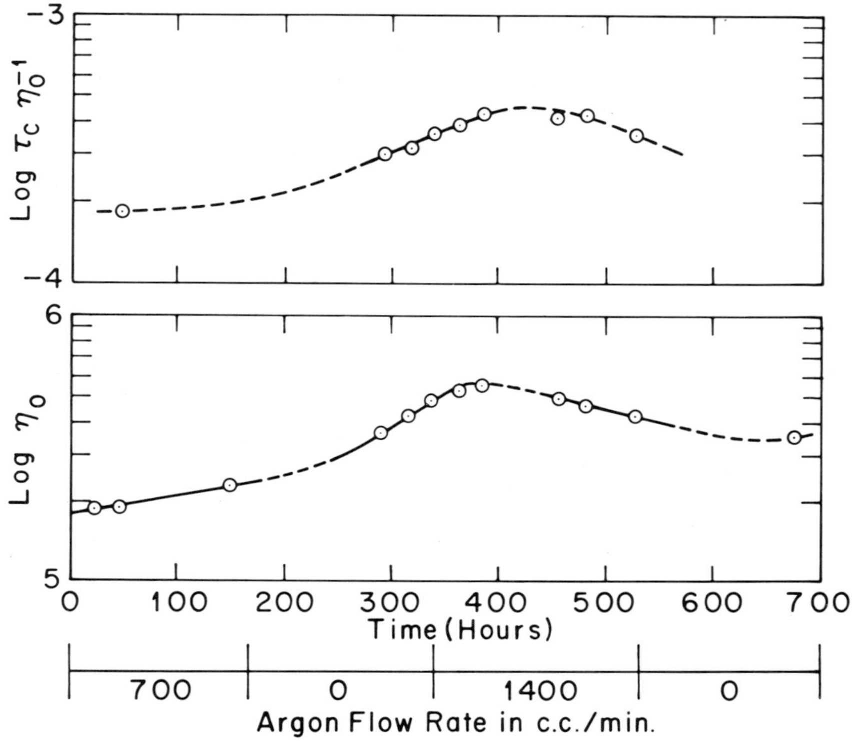

Fig. 8. Change of zero shear viscosity η_o and $\tau_c \eta_o^{-1}$ for the BBB/MSA solution at room temperature as a function of the aging time t. The argon flow rates was varied as indicated on the figure.

Even though the association of BBB in MSA is accompanied by a change in η_κ and the recovery $\gamma_{R,\kappa}$ as functions of shear rate it can be seen in Fig. 9 that the reduced curves η_κ/η_o and R_κ/R_o versus $\tau_c\kappa$ coincide. Here the recovery function R_κ is defined as $\gamma_{R,\kappa}/\eta_\kappa\kappa$. For the data in Fig. 9, η_o was increased by nearly 100-fold in going from the dry solution to the sample with the most amount of water. The superposition of the data in Fig. 9 show that during the early stages of aggregation BBB does not change its basic coil conformation, but the chains become effectively random-branched by the intermolecular aggregation. On this basis relations for η_o and R_o can be utilized to estimate the degree of association. For example, the increase in η_o over the viscosity η_{od} of the

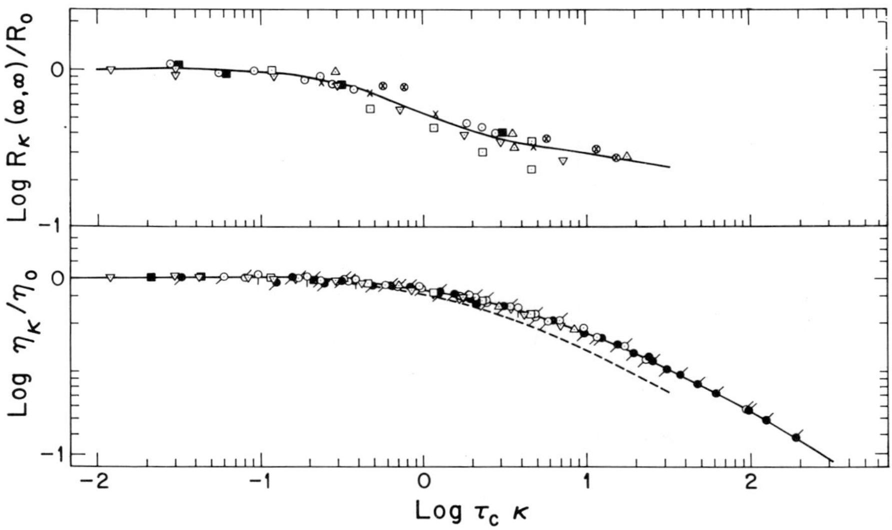

Fig. 9. Reduced logarithmic plot of R_K/R_O and η_K/η_O versus $\tau_C K$. Dashed line, polyisobutylene in cetane with normal molecular weight distribution.

unaggregated polymer (at the same temperature) can be approximated by the relation (14)

$$(\eta_o/\eta_{od})_T = [g(1 + n)]^{3.4} \tag{3}$$

where g is the ratio of the radii of gyration of branched and linear chains of the same molecular weight and n is the number of branch nodes. Assuming that the aggregates have the form of a randomly branched polymer, Eq. (3) can be rewritten using the relation between g and n given by Zimm and Stockmayer (16):

$$(\eta_o/\eta_{od})_T = \left[\frac{1 + n_w}{n_w} \ln(1 + n_w)\right]^{3.4} \approx (1 + n_w)^{1.24} \quad (4)$$

where n_w is the weight average number of branches per molecule (the approximation form is useful for $n_w < 12$). Use of Eq. (4) gives $n_w = 34$ for the sample with the maximum increase in η_o. Similarly, as shown in Fig. 10, the effect

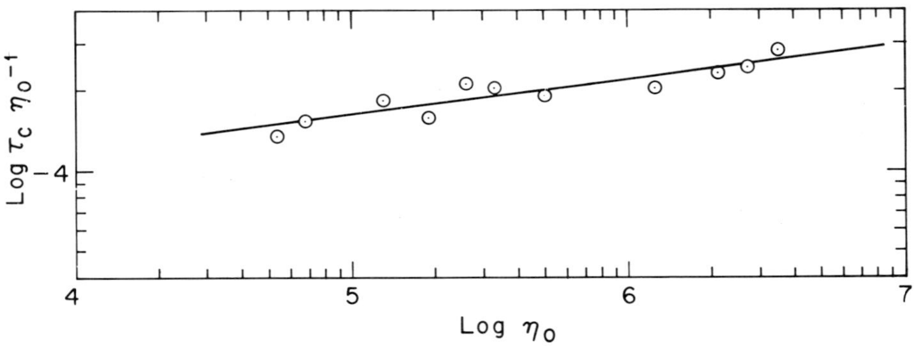

Fig. 10. Logarithmic plot of R_o as given by $\tau_c\eta_o^{-1}$ versus zero shear viscosity η_o.

of aggregation on R_o can be represented by the relation

$$R_o/R_{od} = (\eta_o/\eta_{od})_T^{0.14} \approx (1 + n_w)^{0.175} \quad (5)$$

In fact, since R_o is much less sensitive to temperature than η_o (in general, $R_o \propto 1/T$, Eq. (5)) may be generally more useful in assessing the effect of temperature on aggregation through experimental measurement of R_o than the corresponding use of Eq. (4) with data on η_o.

An important point to be realized from the studies on BBB is that the presence of aggregation would have been very difficult to detect for this coil-like polymer given only the flow curve η_κ versus κ since in many respects this curve is quite normal in behavior. The situation is quite different with the rodlike polymer BBL. As shown in Fig. 11, η_κ is found to increase at small κ, with η_κ becoming nearly inversely proportional to κ so that the steady-state shear σ_{ss} is nearly independent of κ. We believe this is caused by an association producing the brush-heap aggregate struc-

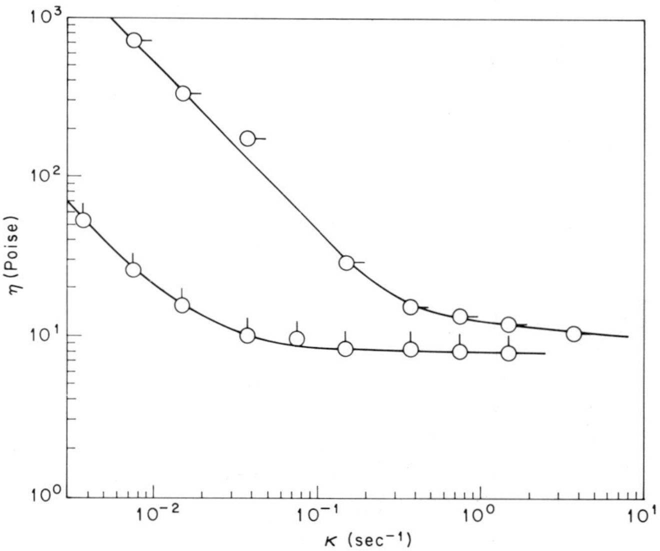

Fig. 11. Logarithmic plot of steady state viscosity η *against shear rate* κ *for a "dry" BBL solutions in MSA,* ⦵ *; and the same solution with 0.5% water,* O- *. (conc = 3.09 g/dl, temp. = 14°C).*

ture described above in relation to association in dilute solutions of BBL. It is this aggregate structure which is continuously broken down and reformed at slow flow conditions. When the shear rate is such that the chains no longer have time to reform the aggregate structure the viscosity becomes independent of κ--eventually, when $\eta_o R_o \kappa$ is of the order unity, the viscosity will again decrease as the rodlike polymer becomes oriented in the flow field.

With higher degrees of aggregation, BBL solutions form weak viscoelastic solids that exhibit plastic flow if the total shear strain exceeds a critical value γ_Y. Some torsional creep and recovery data on such a BBL 'solution' are given in Fig. 12. In this case, the data obey Eq. (1)

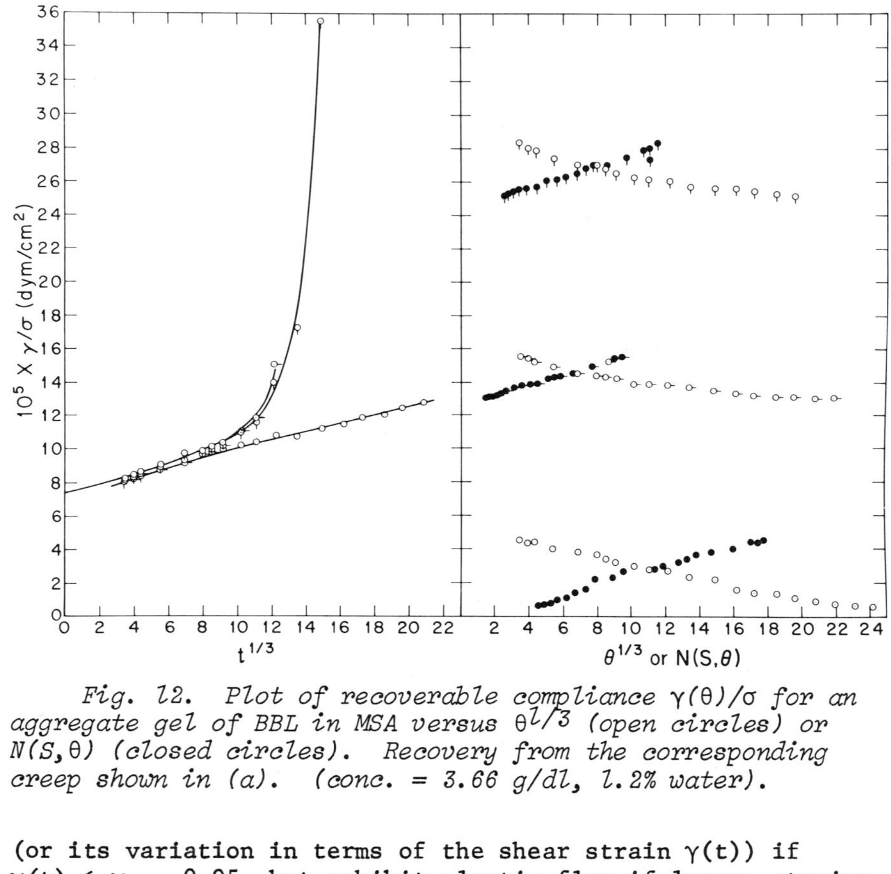

Fig. 12. Plot of recoverable compliance $\gamma(\theta)/\sigma$ for an aggregate gel of BBL in MSA versus $\theta^{1/3}$ (open circles) or $N(S,\theta)$ (closed circles). Recovery from the corresponding creep shown in (a). (conc. = 3.66 g/dl, 1.2% water).

(or its variation in terms of the shear strain $\gamma(t)$) if $\gamma(t) < \gamma_Y \sim 0.05$, but exhibit plastic flow if larger strains are imposed. It may be noted that when plastic flow does occur the recovery can be calculated with Eq. (1) if the nonrecoverable deformation is considered to be viscous-like in its effect on the recovery. Thus application of linear viscoelasticity with the cube-root creep law leads to the relation

$$\gamma(S,\theta) = \sigma J_A \tau_A^{-1/3} N(S,\theta) + \gamma_{NR} \tag{6a}$$

$$N(S,\theta) = (S + \theta)^{1/3} - \theta^{1/3} \tag{6b}$$

where $\gamma(S,\theta)$ is the recovery after recoverying for time θ following creep for time S and γ_{NR} is the nonrecoverable strain (the plastic flow). The filled circles in Fig. 12 giving $\gamma(S,\theta)$ versus $N(S,\theta)$ show that the time constant τ_A for the recoverable deformation is unaffected by the plastic flow that occurs for $\gamma(t) > \gamma_Y$.

V. CONCLUSION

The solubility of the heterocyclic polymers BBB and BBL occurs only when the graphite-like aggregate structure is disrupted by rather specific polymer-solvent interactions. Even though this interaction involves only a modest energy, it is very effective in creating meta-stable aggregates that appreciably alter the properties of the polymers in dilute and concentrated solution and is the dominant factor with regard to the properties of the polymers in the solid state. Protonation occuring with strong protic acids is an example of a specific polymer-solvent interaction that can lead to solubility of BBB and BBL. Interaction with Lewis acids such as antimony trichloride is another example. In the former case, a polyelectrolyte is formed, favoring disruption of interchain aggregates if the polyelectrolyte is compatible with the solvent.

Acknowledgement. The work reported herein has been supported in part by the Air Force Materials Laboratory, Polymer Branch, Wright-Patterson Air Force Base, OH.

VI. REFERENCES

1. Van Deusen, R. L., J. Polym. Sci. Pt. B 4, 211 (1966).
2. Arnold, F. E. and Van Deusen, R. L., J. Polym. Sci., 15, 2035 (1971).
3. Ammons, V. G. and Berry, G. C., J. Polym. Sci., Pt. A-2, 10, 449 (1972).
4. Berry, G. C., Discussions Faraday Soc., 49, 121 (1970).
5. Wong, C. P. and Berry, G. C., Org. Coatings Plast., 33(1) 215 (1973).
6. Berlin, A. A., Liogonkii, B. S., Shamrayen, G. M., and Belova, G. V., Polym. Sci., USSR, 9, 1284 (1967).

7. Berry, G. C. and Eisaman, P., J. Polym. Sci., Phys. Ed. 12, 2253 (1974).
8. Berry, G. C. and Liwak, S. M., J. Polym. Sci., Phys. Ed. in press.
9. Berry, G. C., J. Polym. Sci., Phys. Ed., 14, 479 (1976).
10. Berry, G. C., J. Polym. Sci., Phys. Ed., in press.
11. Gillham, J. K. and Glazier, K. C., J. Appl. Polym. Sci., 16, 2153 (1972).
12. Powell, J. W. and Chartoff, R. P., J. Appl. Polym. Sci., 51, 261 (1968).
13. Berry, G. C. and Wong, C. P., J. Polym. Sci., Phys. Ed. 13, 1761 (1975).
14. Berry, G. C. and Fox, T. G, Adv. Polym. Sci., 51, 261 (1968).
15. Berry, G. C. and Yen, S. P., Adv. in Chemistry Series, 91, 734 (1969).
16. Zimm, B. H. and Stockmayer, W., J. Chem. Phys., 17, 1302 (1949).

VISCOSITIES OF POLYSTYRENE SOLUTIONS

by
José M. Sosa
Chemistry Department
New Mexico State University

and

Arturo Rubio and José Pérez B.[1]
Depto. de Química, Universidad Católica Andrés Bello,
Caracas, Venezuela

I. INTRODUCTION

The dependence of the limiting viscosity number [n] (intrinsic viscosity) of a polymer solution on molecular weight, temperature and solvent has been treated theoretically and confirmed experimentally for several polymers[2,3]. The well-known Mark-Houwink-Sakaruda equation, $[n] = K'M^a$, relates the limiting viscosity to molecular weight (M), by the constants K' and a. From tabulated values of K' and a, a qualitative relation between a and solvent-type can be observed. The value of a tends toward 0.50 for "poor" solvents and toward 0.8 for "good" solvents for most flexible polymers. At the theta (θ) condition, when the polymer chain is in its unperturbed dimension, a is equal to 0.5.

Polymer-solvent interactions are considered to affect the real dimensions of a polymer chain by an expansion factor[4] α, i.e., $[n] = K M^{\frac{1}{2}} \alpha^3$. α has a value of 1.0 at θ conditions and will increase as the solvent swells polymer molecules better. While K is independent of molecular weight and solvent type for linear, flexible polymers having molecular weights over 50,000, the constant α depends on temperature, solvent-type and molecular weight range. Polymer-solvent interactions are governed by the enthalpy and entropy of mixing. Flory[5] utilized the terms "geometrically good" or "geometrically poor" to classify solvents in relation to the structure of the polymer segment. Solvents were also classified as to "thermally good" or "thermally poor". Both of these parameters are important in classifying solvents.

In this paper, the influence of solvents on the limiting viscosity number of polystyrene solutions is explored. The concept of solubility parameters (δ) is used to classify solvents because of its general acceptance and its convenience. Attempts in our laboratory to construct universal calibration curves for molecular weight determination by the gpc technique in part lead to this study. Literature values of K' and a for different polymers were needed to calculate hydrodynamic volumes, and usually several widely differing values were found. The decision of what values to use is not always straight forward, and differences in molecular weights in excess of 100% could be obtained depending on what pair of values were chosen. Recently, Munk and Halbrook[6], point out that in their attempt to use K' and a values to calculate coil dimensions of polymer molecules the problem of selecting the best values was experienced. These authors selected values for polystyrene by evaluating several sets of data, but for other polymers the task was not simple.

A. The Solubility Parameter Concept

The spontaneity of a process is governed by the well-known free energy expression $\Delta G = \Delta H - T\Delta S$. In dissolving a polymer in a solvent, the change in the entropy of mixing (ΔS_{mix}) is always positive. Thus, the enthalpy of mixing (ΔH_{mix}) is the deciding factor as to whether the free energy change ΔH_{mix} is negative. Hildebrand[7] suggested that:

$$\Delta H_{mix} = Vm\left[\left(\frac{\Delta E_1}{V_1}\right)^{1/2} - \left(\frac{\Delta E_2}{V_2}\right)^{1/2}\right]^2 \phi_1 \phi_2,$$

where Vm is the total volume of the mixture, ΔE is the energy of vaporization of component 1 or 2, V is the molar volume of component 1 or 2, and ϕ is the volume fraction of component 1 or 2 in the mixture. $\Delta E/V$ is the energy necessary to vaporize a cc of the component. It can be seen that if $\Delta E/V$ for the components 1 and 2 are equal, then ΔH_{mix} is zero and the entropy term would insure the spontaneity of the process. The solubility parameter (δ) is defined as $(\Delta E/V)^{1/2}$. Thus, if solubility parameters can be calculated for solute and solvent, the solubility of the solute can be predicted. Various methods have been shown to yield acceptable values of δ for solvents [8,9,10]. δ for polymers are estimated using Small's[11] table of molar attraction constants or by noting the degree of swelling of the polymer in different solvents. The solubility parameter of the polymer is similar to that of the solvent where the largest amount of swelling takes place.

B. Effect of Solvents on the Limiting Viscosity Number

The influence of solvents on the limiting viscosity number is shown in figure 1 by a plot of [n] vs δ for polyvinyl acetate. [n] values were calculated from literature values of K' and a by use of the Mark-Houwink-Sakaruda equation. A molecular weight of 100,000 was chosen for the calculation. Thus, in a typical example: $K' = 41.5 \times 10^{-5}$, a = 0.62 at 30°C in acetonitrile (δ = 11.9) and $[n] = 41.5 \times 10^{-5}(10^5)^{0,62} = 0.522$ dl/g. This curve is typical for polymers and is expected from qualitative considerations of the expansion of a polymer chain in a poor or good solvent.

As the solubility parameter of the solvent approaches that of the polymer, the limiting viscosity number increases.

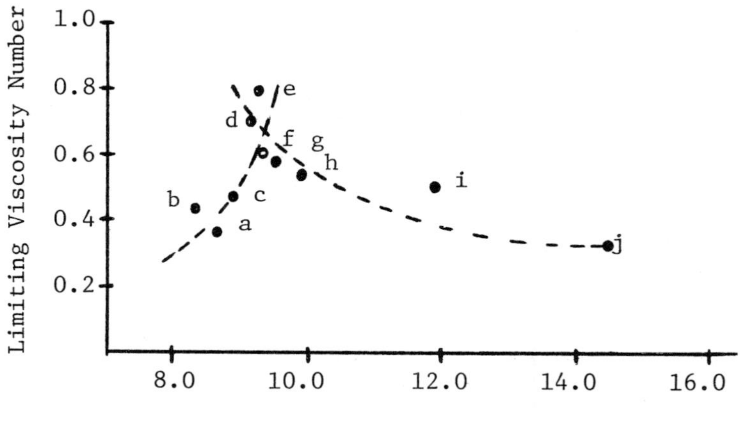

Figure 1. Plot of limiting viscosity number vs solubility paramater for poly(vinyl acetate). $[n] = K'(10^5)^a$, where K and a are literature values. a-butanone, b-methylisobutyl ketone, c-toluene, d-benzene, e-chloroform, f-ethylformate, g-chlorobenzene, h-acetone, i-acetonitrile, j-methanol.

K' and a values used in this calculation were obtained by different investigators. Light scattering or osmometry were used to characterize the fractions that in turn were used to evaluate K' and a. Similar curves are obtained for polystyrene, poly(methyl methacrylate) and polyisobutene. Data for other polymers are not sufficient.

Figure 2 shows a plot of -log [n] vs log $|\delta_p - \delta_s|$ for polyvinyl acetate using the same values as for figure 1, where δ_p and δ_s are the solubility parameters for polymer and the solvent, respectively. The solubility parameter for the polymer (δ_p) is obtained by estimating the value of δ_s that gives the highest [n] value from a graph of [n] vs δ_s. An excellent correlation exists between - log [n] and log $|\delta_p - \delta_s|$ for the different solvents. The linear correlation coefficient (r) is equal to 0.934 for 11 points giving over 99.9% confidence that a relationship does exist. Similar plots for polystyrene, poly(methyl methacrylate) and polyisobutene are also shown.

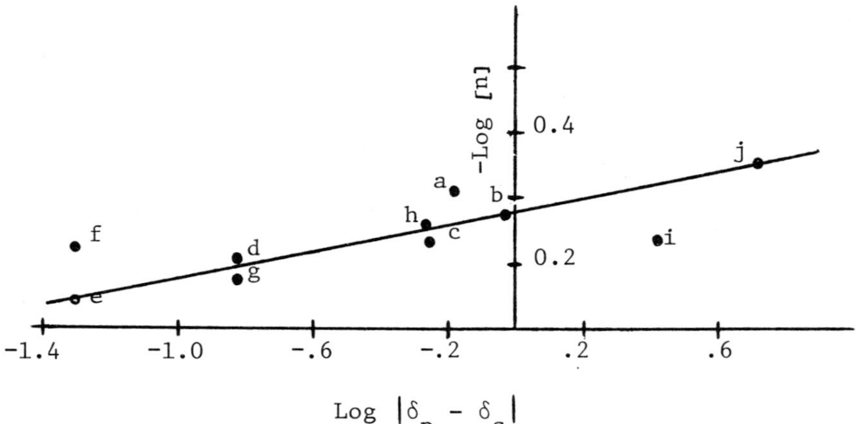

Figure 2. Plot of -log of the limiting viscosity number [n] vs log of the absolute value of the difference between the solubility parameter of the polymer δ_p and of the solvent δ_s for poly(vinyl acetate) ($\delta_p = 9.35$). Solvent legend is the same as for Figure 1. The linear correlation coefficient (r) is 0.934 for seven points omitting points a, f, and i.

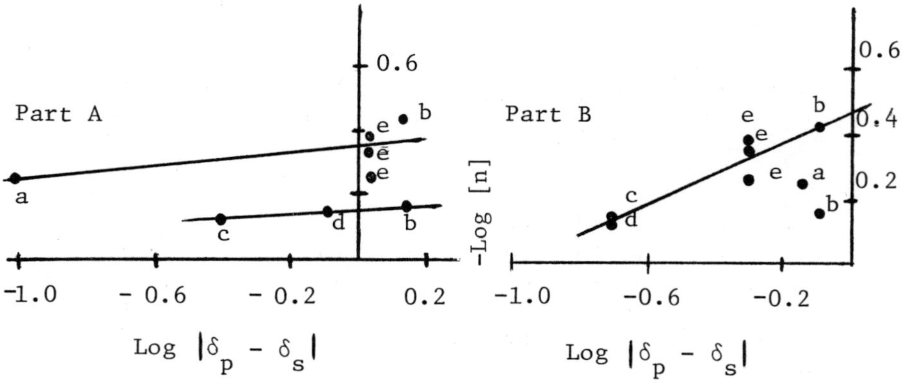

Figure 3. Plot of -log of the limiting viscosity number vs log of the absolute value of the difference between the solubility parameter of the polymer (δ_p) and the solvent (δ_s) for polyisobutene. [n] = $K'(10^5)^a$, where K' and a are literature values. Solvents: a-diisobutylene, b-benzene (2 values), c-cyclohexane, d-carbon tetrachloride, e-toluene (3 values).
Part A. δ_p = 7.8 from the literature. Part B. δ_p = 8.4 estimated from plot of [n] vs δ_s.

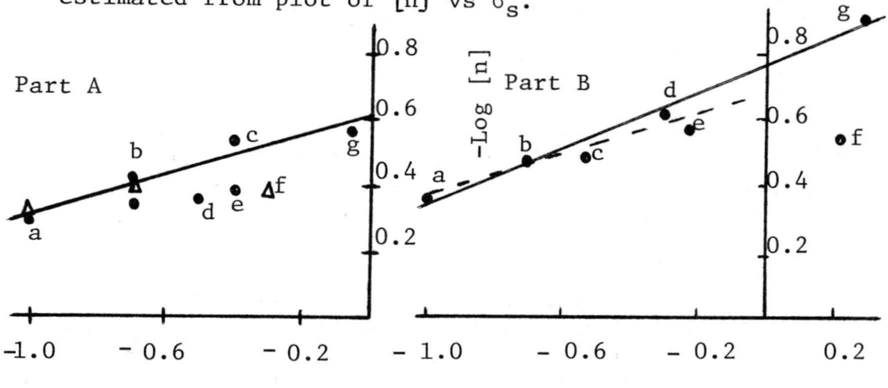

Figure 4A. Plot of -log [n] vs log $|\delta_p - \delta_s|$ for polystyrene (δ_p = 9.1). [n] = $K'(10^5)^a$, where K' and a are literature values. Solvent key: a-benzene (2 values), b-toluene (3 values), c-butanone (2 values), d-ethylbenzene, e-chlorobenzene, f-carbon tetrachloride, g-cyclohexane, Δ indicate experimental values obtained in this work. Figure 4B. Plot of -log [n] vs log $|\delta_p - \delta_s|$ for poly(methyl methacrylate) (δ_p = 9.4). Solvent key: a-chloroform, b-benzene, c-ethyl acetate, d-acetone, e-toluene, f-nitroethane, g-acetonitrile.

The relationship observed between -log [n] and log $|\delta_p-\delta_s|$ is sensitive to the value of δ_p that is selected. This is shown for polyisobutene in figure 3. If a literature value of 7.8 for δ_p is selected, the scatter of points is such that apparently two distinct curves can be drawn thru the set of points. However, if δ_p is estimated from a plot of [n] vs δ, only two of the eight points do not correlate well. Thus, these points may be representative of different samples or perhaps they are incorrect. While this empirical relationship must be tested with a wider number of polymers in different solvents, it does help in separating those values that for one reason or other deviate from the trend established by the other values.

A molecular weight of 100,000 was used to calculate the limiting viscosity numbers since most K',a values are considered to be valid in the 10^4-10^6 molecular weight range. If molecular weights of 10^3, 10^4, 10^5 and 10^6 are used to calculate the limiting viscosity numbers and if the calculated values are used in -log [n] vs log $|\delta_p-\delta_s|$ plots, it is found that the points tend to scatter widely as lower molecular weights are approached. By changing the value of δ_p, the correlations are improved somewhat, but the nature of this scatter is difficult to explain at this time.

The ability to predict values for K' and a for different solvents would perhaps be the ultimate test. This could be done by assuming a value of K' and determining a by referring to a -log [n] vs log $|\delta_p-\delta_s|$ plot at a given molecular weight.

Since polystyrene standards are readily available, it was decided to test the relationship with experimentally determined values. This paper presents preliminary results obtained in three solvents with narrow distribution polystyrene samples and for a commercial sample of polystyrene in seven solvents.

II. EXPERIMENTAL

Five polystyrene standards having molecular weights of 35,000, 111,000, 200,000, 498,000 and 650,000 obtained from Waters Associates were used in this study. Benzene, carbon tetrachloride and toluene (Fisher Reagent Grade) were chosen as the solvents. Limiting viscosity numbers for the five standards in benzene were obtained by extrapolating n_{sp}/C vs C plots to zero concentration. Limiting viscosity numbers for the standards in carbon tetrachloride and toluene were obtained by the one point method. All measurements were made at 30°C using an Ubbelode viscometer that had a discharge time of 114.0 seg for benzene at this temperature. Relative viscosities were determined for the commercial sample in the solvents chosen. Shell crystal grade PS having a moleculor weight ($\bar{M}n$) of 140,000 and a $\bar{M}w/\bar{M}n$ of 2.3 as determined by gpc was used.

III. RESULTS AND DISCUSSION

Figure 5 shows a plot of log [n] vs log M for the five samples in the three solvents. The following K′ and a values were obtained at 30°C:

	K′	a
Benzene	15.4×10^{-5}	0.70
Toluene	15.0×10^{-5}	0.69
CCl_4	7.24×10^{-5}	0.75

The experimental values of [n] obtained in toluene and benzene from log [n] vs log M plots fall on the best curve drawn on the plot - log [n] vs log $|\delta_p - \delta_s|$ for polystyrene as shown on figure 4A by the triangles. The same was not true for the [n] value in CCl_4 which falls close to that of the literature value of chlorobenzene. A value of 0.73 for a instead of 0.75 would account for the deviation if the K′ value is assumed to be 7.24×10^{-5}. On the other hand if a value of 15×10^{-5} is selected for K′, a would be 0.664. The value of 0.664 is more reasonable based on the qualitative reasoning utilized to explain a values. However, presently, no explanation can be given for the relatively high value 0.75 obtained experimentally. CCl_4 is apparently a much better solvent than what its solubility parameter would indicate, and it may be that other factors must be considered to explain this fact. It should be remembered that the entropy of mixing is being ignored, since the solubility parameter relationship deals basically with the enthalpy of mixing.

In order to test the proposed relationship further, relative viscosities (n_{sp}) of polystyrene in seven aromatic solvents (0.5 gms/100 ml of solution) were determined. A plot of log n_{sp} vs log $|\delta_p - \delta_s|$ shows that with the exception of ethylbenzene and toluene the points fall on a straight line. (Fig. 6)

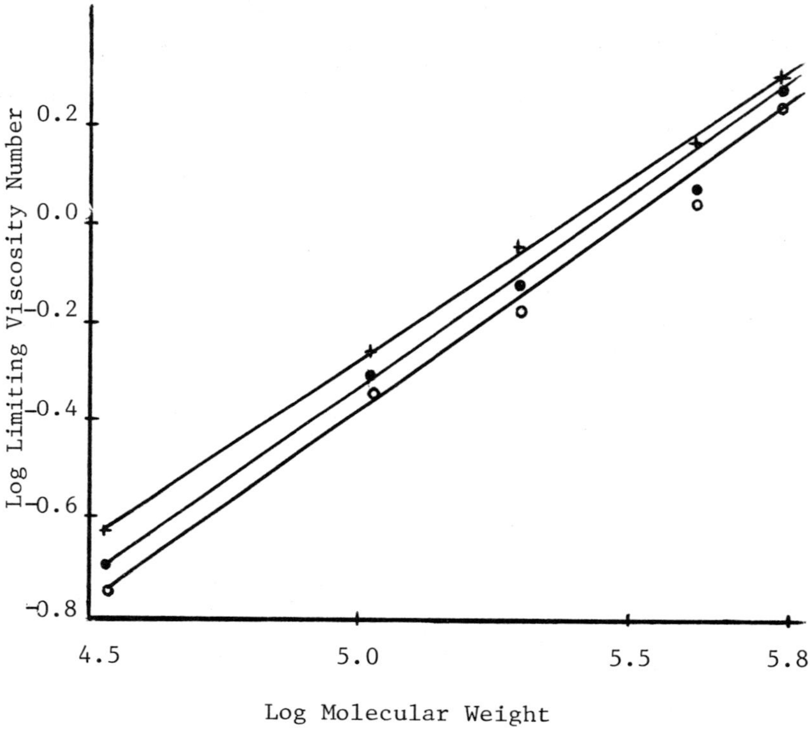

Figure 5. Plot of log of the limiting viscosity number vs log molecular weight for polystyrene standards. + benzene, toluene, o CCl_4

Solubility parameters, appear to correlate fairly well the dilute solution viscosities of several polymers. Exactly what contribution this can make to the complex theory of liquids and solutions is not clear. On the practical side, if K' values were standardized, values of a for different solvents could be obtained by determining the relative viscosity of the polymer, obtaining log N_{sp} vs log $|\delta_p - \delta_s|$ plots and simply reading the a value from the graph. In essence, what appears to follow from the empirical relationship is that it may be possible to obtain expansion factors for a polymer by using δ values.

IV. CONCLUSIONS

A relationship between limiting viscosity number [n] and solubility parameters has been obtained using literature values of K' and a. It is found that plots of log [n] vs log $|\delta_p - \delta_s|$, where δ_p and δ_s are the solubility parameters of the polymer and the solvent, respectively, give a linear relationship. This relationship works well for polystyrene, poly(vinyl acetate), poly(methyl methacrylate) and poly isobutene. Values of δ_p can be obtained by plotting [n] vs δ_s and reading the δ_s that gives the highest [n]. Experimental [n] values for polystyrene in benzene and toluene correlate well, however, the value for CCl_4 does not. Presently, work is being carried out with polystyrene to test the relationship with other solvents. Particular attention is being given to those solvents that show deviations, hoping that a sounder basis can be found for this relationship.

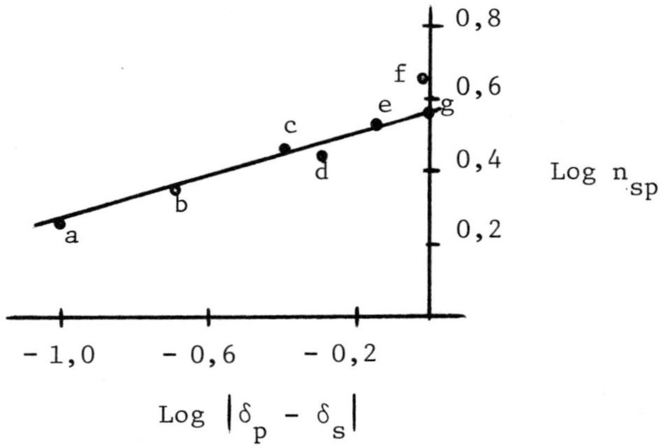

Figure 6. Plot of -log n_{sp} vs log $|\delta_p - \delta_s|$ for polystyrene (δ_p estimated at 9.3). Solvent key: a-benzene, b-chlorobenzene,c-toluene, d-ethylbenzene, e-nitrobenzene, f-benzonitrile, g-aniline.

V. REFERENCES

1. Taken in part from the work presented by J.P.B. for his licenciatura degree.
2. P.J. Flory and T.G. Fox, Jr., JACS, 73, 1904 (1951).
3. W.R. Moore, "Viscosities of Dilute Polymer Solutions" in Progress in Polymer Science, A.D. Jenkins, Editor, Pergamon Press, New York, 1, (1967).
4. P.J. Flory, Principles of Polymer Chemistry, Cornell University Press, Ithaca, N.Y., 1953.
5. P.J. Flory and T.G. Fox, JACS, 73, 1909, 1951.
6. P. Munk and M.E. Halbrook, Macromolecules 9 (3), 441 (1976).
7. J. Hildebrand and R. Scott, The Solubility of Non-electrolytes, 3rd Edition, Reinhold Publishing Corp., New York, 1949.
8. J. Brandrup and E.A. Immergut, Editors, Polymer Handbook, Interscience Publishers, New York, 1966.
9. R.B. Seymour and J.M. Sosa, Nature, 248, (5451), 759 (1974).
10. R.B. Seymour, Aust. Paint J., 13, (10), 18 (1968).
11. P.A. Small, J. Appl. Chem., 3, 71 (1953).

DETERMINATION OF SOLUBILITY PARAMETERS OF COPOLYMERS OF STYRENE AND ACRYLONITRILE

Raymond B. Seymour and Hubert A. Wood
Department of Polymer Science
University of Southern Mississippi

I. INTRODUCTION

The solubility of poly (Styrene-co-acrylonitrile) differs from that of polystrene (δ = 9.3H) and polyacrylonitrile (δ = 12.5H) and varies with the composition of the copolymer. For example, a copolymer consisting of 54 and 46 percent of styrene and acrylonitrile, respectively, is insoluble in benzene (δ = 9.15H), toluene (δ = 8.91H), and butyl acetate (δ = 8.46H), but is soluble in acetone (δ = 9.77H) which is a nonsolvent for both homopolymers.

Solubility parameters may be calculated by an extension of procedures developed previously(1) using three dimensional solubility parameter concepts(2).

II. EXPERIMENTAL

The techniques used may be demonstrated by that employed for the investigation of a copolymer with an average molecular weight of 270 thousand and consisting of 54 and 46 percent of styrene and acrylonitrile, respectively. A dilute solution of this copolymer was titrated slowly with a nonsolvent and the end point or cloud point was determined. The original concentration of the solution was such that approximately 40 mg of copolymer was present in 100 ml of mixed solvent at the end point.

III. DISCUSSION

The Hansen (2)(3)(4) three dimensional solubility parameter system is based on the concept that internal pressure is the sum of three independent interactions. These interactions are the result of dispersion forces, polar forces, and hydrogen bonding forces and are designated by a subscript d, p, or h, respectively. Thus since Hildebrand(5) defined the solubility parameter as the square root of the internal pressure, the Hansen parameters are defined as follows:

$$\delta_d = \frac{\Delta E_d^v}{V_m}$$

$$\delta_p = \frac{\Delta E_p^v}{V_m}$$

$$\delta_h = \frac{\Delta E_h^v}{V_m}$$

Where δ is the solubility parameter, ΔE^v is the energy of vaporization and V_m is the molar volume.

These three parameters are related to the overall or Hildebrand solubility parameter by the following equation:

$$\delta_o^2 = \delta_d^2 + \delta_p^2 + \delta_h^2$$

Solubility of a polymer is represented in the three dimensional space of the Hansen parameters as a spheroid. The location of the center of the spheroid is the absolute solubility parameter value for the polymer. The radius of the spheroid is the plus and minus range, $\Delta\delta_o$, for complete miscibility of the polymer[1]. Deviations from solution ideality or interaction between separated solubility components may alter the geometric shape of the solubility zone. Deviations should become evident experimently.

Once outside of the critical solubility limits, polymer solubility should decrease rapidly. It is in this range that the Flory, Huggins interaction parameter μ can be used to evaluate polymer solubility parameters. Suh and Clarke(1) used the following equation:

$$\mu = \frac{1}{Z} + \frac{V_1 (\delta_1 - \delta_3)^2}{RT}$$

---

[1] Pages 371-2, Chapter XX of Hildebrand's The Solubility of Nonelectrolytes give a good account of the solubility range and critical value μ_c.

Where Z is the coordination number, V_1 is the molar volume of the solvent and δ_1 and δ_3 are the solubility parameters of the solvent and polymer, respectively.

For dilute solutions of high polymers at or above the critical solubility limits μ tends to approach a constant value(6). Thus for the single solubility parameter, it should be possible to approach the value δ_3 from opposite sides of the solubility spectrum and equate the two resulting equations. This is best done by titrating a dilute solution of polymer in a good solvent with more and less polar non-solvents until each causes a polymer phase to precipitate.

$$\mu_l = \left(\frac{1}{Z}\right)_l + \frac{V_{ml}(\delta_3 - \delta_{ml})_l^2}{RT}$$

$$\mu_h = \left(\frac{1}{Z}\right)_h + \frac{V_{mh}(\delta_{mh} - \delta_3)_h^2}{RT}$$

Subscripts l and h represent low and high solubility parameter respectively.

μ_l and μ_h can be equated and the resulting equation can be further reduced by considering the entropy terms, 1/Z, approximately equal and removing them also.

$$\sqrt{V_{ml}}\ (\delta_3 - \delta_{ml})_l = \sqrt{V_{mh}}\ (\delta_{mh} - \delta_3)_h$$

Solving for δ_3 yields the following:

$$\delta_3 = \frac{\sqrt{V_{ml}}\ \delta_{ml} + \sqrt{V_{mh}}\ \delta_{mh}}{\sqrt{V_{ml}} + \sqrt{V_{mh}}}$$

Values of δ_3 are best for situations where the solvent-polymer interaction is small compared to the solvent-nonsolvent interaction. This is seldom the case as:

$$(\delta_{mh} - \delta_1) \cong (\delta_{mh} - \delta_3)$$

or

$$(\delta_1 - \delta_{ml}) \cong (\delta_3 - \delta_{ml})$$

When the solvent-polymer interactions become substantial enough to compete with solvent-nonsolvent interactions, then δ_m will be consistently the same distance from δ_3. That is, the plot of δ_m values will produce a line parallel to the $\delta_1 = \delta_m$ line in a graphical representation of δ_1 vs. δ_m.

By substituting either of the above two equations into the original equation for δ_3, one can generate a single equation to account for these associations.

$$\delta_3{*} = \delta_{ml} + \sqrt{(V_{mh}/V_{ml})} \quad (\delta_{mh} - \delta_1)$$

$\delta_3{*}$ may equal δ_3 when both equations are equally bad. In either case, the values for solubility parameter should be obtained by extropolating to zero nonsolvent concentration, that is where $\delta_3 = \delta_1$.

The resulting values for V_m and δ_m can be determined using Scott's "single liquid" approximation.

$$V_m = \frac{V_1 V_2}{\phi_1 V_2 + \phi_2 V_1}$$

$$\delta_m = \phi_1 \delta_1 + \phi_2 \delta_2$$

Where ϕ is the molar volume fraction.

Since each of the Hansen parameters function independently, then each should be soluble separately with the above technique. The first requirement is that each parameter be examined separately without manipulating the other two. This can be accomplished if solvents and nonsolvents are chosen such that only one parameter varies greatly. The second requirement is that each titration must approach the parameter from opposite ends of the solvent spectrum in order that a δ_{ml} and δ_{mh} may be evaluated. This means that for each parameter, there must be one nonsolvent with a parameter value above that of the polymer and one below. As will be seen, these are not always easily met.

Table 1 shows the solubility parameter data for the solvents and nonsolvents while Table 2 shows the experimental data for the turbidimetric titrations. A samply of the preliminary data for the overall solubility parameter δ_o from n-hexane and methanol titrations is shown in Figure 1. It can be seen that the methanol titrations parallel the $\delta_1 = \delta_m$ line indicating strong solvent-polymer interaction as compared to solvent-nonsolvent interaction.

TABLE 1

Solubility Parameter Values of Solvents and Nonsolvents (8)

	NAME	Solubility Parameter δ_o	δ_d	δ_p	δ_h	Molar volume v
1.	methyl isobutyl ketone	8.57	7.49	3.0	2.0	125.8
2.	ethyl acetate	9.10	7.44	2.6	4.5	98.5
3.	methyl acetate	9.15	7.60	3.5	3.7	90.1
4.	mesityl oxide	9.20	7.97	3.5	3.0	115.6
5.	chloroform	9.21	8.65	1.5	2.8	80.7
6.	methyl ethyl ketone	9.27	7.77	4.4	2.5	90.1
7.	tetrahydrofuran	9.52	8.22	2.8	3.9	81.7
8.	chlorobenzene	9.57	9.28	2.1	1.0	102.1
9.	benzonitrile	9.70	8.50	4.4	1.6	102.6
10.	acetone	9.77	7.58	5.1	3.4	74.0
11.	butyronitrile	9.96	7.50	6.1	2.5	87.0
12.	dioxane	10.00	9.30	0.9	3.6	85.7
13.	ethylene dichloride	10.17	9.30	3.6	2.0	79.4
14.	diacetone alcohol	10.18	7.65	4.0	5.3	124.2
15.	benzaldehyde	10.40	9.15	4.2	2.6	101.5
16.	methyl salicylate	10.59	7.80	3.9	6.0	129.0
17.	tetrachloroethane	10.59	9.20	2.5	4.6	80.9
18.	acetophenone	10.63	9.60	4.2	1.8	117.4
19.	pyrodine	10.61	9.25	4.3	2.9	80.9
20.	aniline	11.04	9.53	2.5	5.0	91.5
21.	nitroethane	11.09	7.80	7.6	2.2	71.5
22.	meta-cresol	11.11	8.82	2.5	6.3	104.7
23.	acylonitrile	11.90	7.50	8.8	3.0	67.1
24.	dimethylformamide	12.14	8.52	6.7	5.5	77.0
25.	dimethylsulfoxide	12.93	9.00	8.0	5.0	71.3
26.	hexane	7.24	7.23	0.0	0.0	131.6
27.	methanol	14.28	7.42	6.0	10.9	40.6
28.	carbon disulfide	10.00	10.00	0.0	0.3	60.0
29.	diethyl ether	7.62	7.05	1.4	2.5	104.8

TABLE 2

Composition of Binary mixtures at the cloud point for Turbidimetric Titrations of Poly(styrene 54-co-Acrylonitrile 46)

	Solvent	Volume fraction ϕ_2 of Nonsolvent: hexane	methanol	diethyl ether	carbon disulfide
1.	methyl isobutyl ketone	0.160	0.583	--	--
2.	ethyl acetate	0.129	0.542	--	--
3.	methyl acetate	0.149	0.348	--	--
4.	mesityl oxide	0.302	0.721	--	--
5.	chloroform	0.269	0.644	--	0.750
6.	methyl ethyl ketone	0.251	0.634	--	--
7.	tetrahydrofuran	0.250	0.727	--	0.650
8.	chlorobenzene	0.091	0.669	--	--
9.	benzonitrile	0.390	0.763	--	--
10.	acetone	0.250	0.595	--	0.790
11.	butyronitrile	0.357	0.671	--	--
12.	dioxane	0.182	0.739	--	0.760
13.	ethylene dichloride	0.288	0.669	--	--
14.	diacetone alcohol	0.288	0.534	--	--
15.	benzaldehyde	0.356	0.781	--	--
16.	methyl salicylate	0.271	0.773	--	0.720
17.	tetrachloroethane	0.450	0.644	--	--
18.	acetophenone	0.417	0.806	--	0.820
19.	pyridine	0.334	0.783	--	--
20.	aniline	--	0.785	0.640	--
21.	nitroethane	--	0.716	0.605	--
22.	meta-cresol	0.510	0.856	--	--
23.	acylonitrile	0.280	0.532	--	--
24.	dimethyl formamide	--	0.775	0.690	--
25.	dimethyl sulfoxide	--	0.778	0.526	--

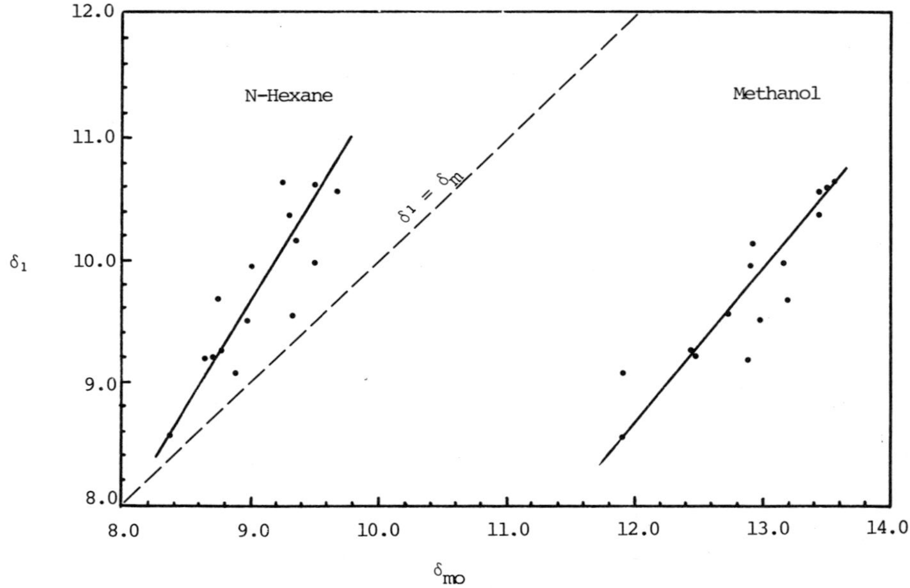

Figure 2 shows the preliminary data for the dispersive solubility parameter for six solvents titrated with hexane, methanal and carbon disulfide. Only, hexane shows any trend toward paralleling the $\delta_1 = \delta_m$ line.

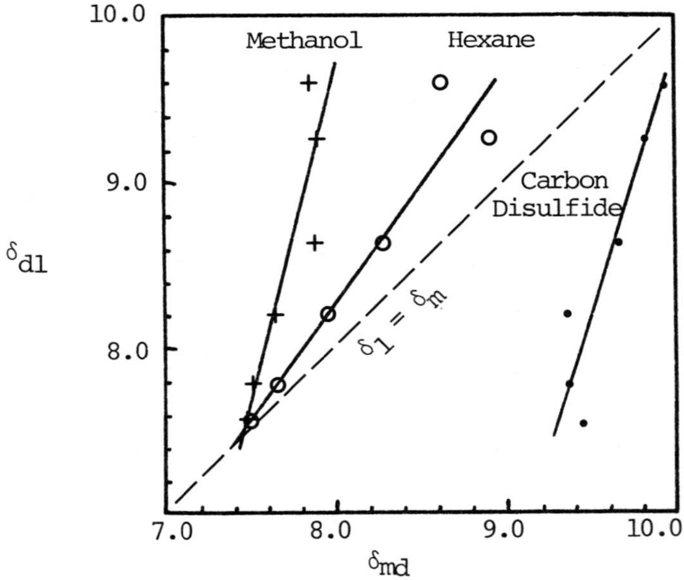

Fig. 2. Turbidimetric titration of Poly (Styrene 54-co-acrylonitrile 46) for six solvents selected for titration with n-hexane, methanol and carbon disulfide.

Table 3 shows the solubility parameters calculated from each solvent by n-hexane and methanol titrations. Values are given for both solvent-polymer; δ_3*, and solvent-nonsolvent, δ_3, interaction. The results can be desceptive since both δ_{d3} and δ^*_{d3} appear to line up around 7.4 or 7.5. In fact, these values are the result of the δ_{d1} of the two non-solvents of 7.2 and 7.4.

TABLE 3

Determination of the Solubility Parameter of Poly (Stryrene 54-co - Acrylonitrile 46) using Hexane and Methanol as Nonsolvents.

		Solubility Parameters							
	Solvent	δ_{o3}	δ^*_{o3}	δ_{d3}	δ^*_{d3}	δ^*_{p3}	δ_{p3}	δ_{h3}	δ^*_{h3}
1.	methyl isobutyl ketone	9.79	10.60	7.45	7.42	3.42	3.69	3.89	5.16
2.	ethyl acetate	10.16	10.95	7.42	7.40	3.19	3.62	5.64	6.49
3.	methyl acetate	9.82	10.37	7.54	7.49	3.61	3.71	4.55	5.26
4.	mesityl oxide	10.29	10.99	7.68	7.49	3.56	3.60	4.68	5.76
5.	chloroform	10.31	11.12	8.10	7.68	2.50	3.25	4.59	5.92
6.	methyl ethyl ketone	10.32	11.07	7.60	7.47	4.18	4.02	4.37	5.72
7.	tetrahydrofuran	10.65	11.47	7.83	7.55	3.37	3.78	5.47	6.61
8.	chlorobenzene	10.74	11.56	8.66	8.22	3.06	3.73	3.67	5.53
9.	benzonitrile	10.51	11.04	7.87	7.47	3.84	3.47	4.02	5.60
10.	acetone	10.59	11.22	7.49	7.42	4.62	4.25	4.87	6.00
11.	butyronitrile	10.60	11.06	7.42	7.36	4.79	3.87	4.31	5.59
12.	dioxane	11.03	11.77	8.50	7.92	2.38	3.44	5.47	6.81
13.	ethylene dichloride	10.85	11.34	8.42	7.79	3.67	3.73	4.17	5.74
14.	diacetone alcohol	10.58	10.85	7.53	7.45	3.75	3.58	5.61	5.82
15.	benzaldehyde	10.93	11.28	8.21	7.59	3.85	3.62	4.60	5.90
16.	methyl salicylate	11.11	11.42	7.60	7.47	3.86	3.83	6.42	6.68
17.	tetrachloroethane	10.70	10.77	8.20	7.49	2.78	2.98	5.08	5.42
18.	acetophenone	10.89	11.05	8.32	7.52	3.67	3.34	4.92	6.85
19.	pyridine	11.14	11.51	8.27	7.58	4.00	3.79	4.91	6.32
20.	meta-cresol	10.94	10.84	7.86	7.27	2.86	3.08	5.82	5.52
21.	acylonitrile	11.75	11.63	7.45	7.40	6.77	5.15	4.40	5.52

In Table 4, data are presented from titrations with n-hexane and carbon disulfide, both of which have essentially no polar or hydrogen bonding forces. Hence, no bracketing of the polymer value could be achieved for these parameters. Figure 3 gives a graphic representation of the extrapolation to zero ϕ where the $\delta_1 = \delta_3$ lines intersect the δ^*_o or δ_o lines. Values for the overall solubility parameters can be read on the graph as 9.25 for δ_{o3} and 9.35 for δ^*_{o3}.

TABLE 4

Determination of the Solubility Parameter of Poly (Styrene 54 - co - Acrylonitrile 46) using Hexane and Carbon Disulfide as Nonsolvents.

	Solvent	Solubility Paramerers							
		δ_{o3}	δ^*_{o3}	δ_{d3}	δ^*_{d3}	δ_{p3}	δ^*_{p3}	δ_{h3}	δ^*_{h3}
1.	acetone	9.52	9.25	8.40	9.11	-	-	-	-
2.	dioxane	9.73	9.10	9.34	9.36	-	-	-	-
3.	chloroform	9.20	9.20	8.91	9.12	-	-	-	-
4.	tetrahydrofuran	9.37	9.20	8.62	8.98	-	-	-	-
5.	acetophenone	9.61	8.21	9.17	8.85	-	-	-	-
6.	methyl salicylate	9.88	8.89	8.38	8.82	-	-	-	-

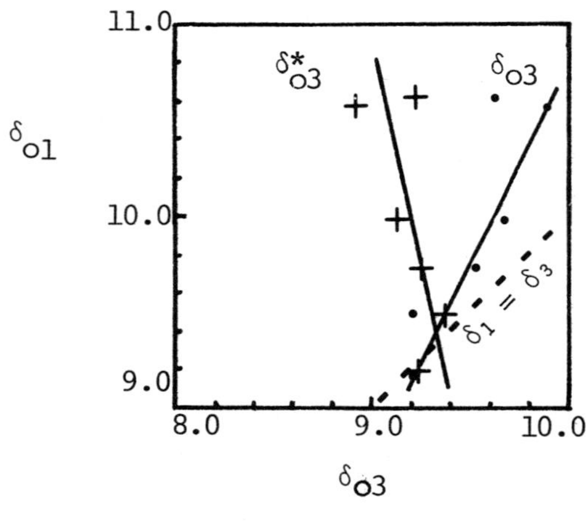

Fig. 3. Determination of the solubility parameters δ_o and δ^*_o for poly (styrene 54-co-Acrylonitrile 46) from the turbidimetric titration with Hexane and carbon disulfide.

Data provided in Table 5 from titrations with methanol and carbon disulfide are graphically interpreted in Figures 4 and 5. Figure 4 evaluates the dispersive solubility parameters δ_{d3} and δ^*_{d3} and both appear to intersect the $\delta_1 = \delta_3$ line at 8.8. Figure 5 evaluates the polar solubility parameter with values of 2.75 and 2.70 for δ_{p3} and δ^*_{p3} respectively.

Table 5
Determination of the Solubility Parameter of Poly(Styrene 54 - co - Acrylonitrile 46) Using Methanol and Carbon Disulfide as Nonsolvents.

Solvent	Solubility Parameters							
	δ_{o_3}	δ_{o_3}	δ_{d_3}	δ_{d_3}	δ_{p_3}	δ_{p_3}	δ_{h_3}	δ_{h_3}
acetone	11.13	12.66	8.53	9.59	3.22	1.55	4.21	4.93
dioxane	11.70	13.16	8.95	8.53	2.27	3.44	4.73	5.70
chloroform	11.05	12.39	8.82	9.01	2.55	3.48	4.25	5.52
tetrahydrofuran	11.26	12.71	8.59	9.03	2.54	2.32	4.94	5.80
acetophenone	11.69	12.58	8.98	8.23	2.99	1.98	4.39	6.58
methyl salicylate	11.65	12.52	8.53	9.42	3.09	2.43	5.47	5.03

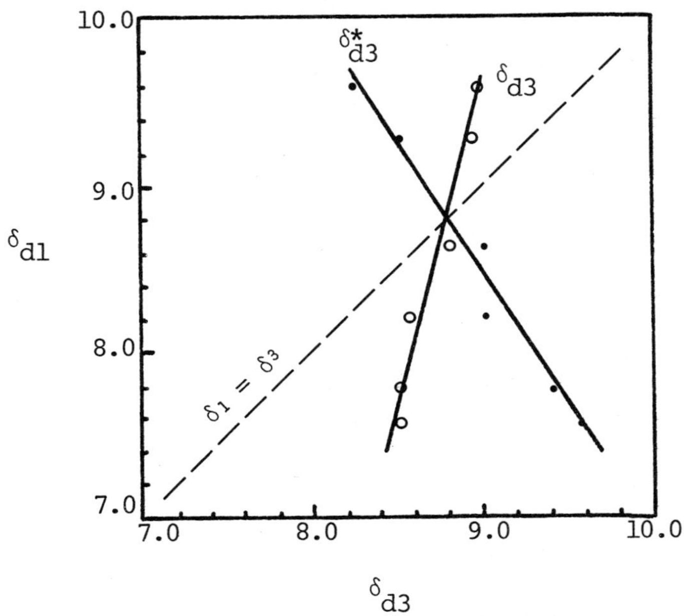

Fig. 4. Determination of the solubility parameters δ_d and δ^*_d for poly (styrene 54-co-Acrylonitrile 46) from the turbimetric titration with methanol and carbon disulfide of Table 5.

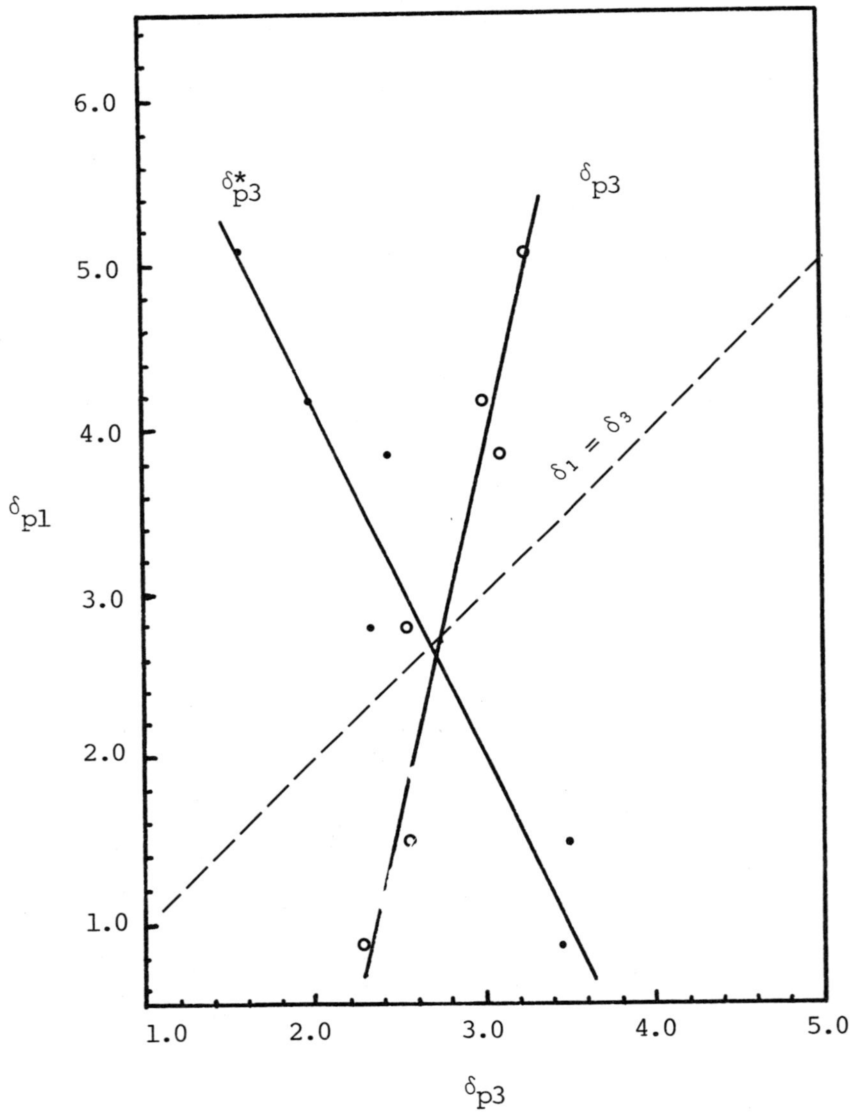

Fig. 5. Determination of the solubility parameters δ_p and δ^*_p for poly (styrene 54-co-Acrylonitrile 46) from turbimetric titration with methanol and carbon disulfide.

Table 6 presents data for a series which were not miscible with hexane in the solubility range required to precipitate the polymer from solution. Individual values are not inconsistent with other individual values of Table 3 for n-hexane titrations. The difference is in the polar and hydrogen bonding parameters of the two nonsolvents Values δ_{d3} and δ^*_{d3} are also locked in as was the case in n-hexane-methanol titrations.

TABLE 6

Determination of the Solubility Parameter Poly (Styrene 54 - co - Acrylonitrile 46) Using Diethyl Ether and Methanol and Nonsolvents.

	Solvent	Solubility Parameters							
		δ_{o3}	δ^*_{o3}	δ_{d3}	δ^*_{d3}	δ_{p3}	δ^*_{p3}	δ_{h3}	δ^*_{h3}
1.	aniline	10.77	10.58	7.91	6.81	3.20	3.67	5.93	6.56
2.	nitroethane	10.82	10.62	7.42	7.15	4.94	3.31	4.92	6.89
3.	dimethylformamide	10.98	10.17	7.58	6.93	4.32	2.66	6.00	6.34
4.	dimethyl sulfoxide	11.75	10.89	7.89	7.08	5.33	3.40	6.17	7.02

Table 7 summarizes the result from all the turbimetric titration experiments. Values have been left out when the data were inconsistant for graphical interpretating.

TABLE 7

Summary: Predicted Solubility, Parameters for Poly (Styrene 54 - co - Acrylonitrile 46) TABLES 3 - 6

	Non-solvent pairs	Solubility Parameters							
		δ_{o3}	δ^*_{o3}	δ_{d3}	δ^*_{d3}	δ_{p3}	δ^*_{p3}	δ_{h3}	δ^*_{h3}
1.	Hexane-methanol	11.10	10.7	7.42	7.32	3.60	3.75	7.0	7.0
2.	Hexane-carbon disulfide	9.25	9.35	9.25	8.92	--	--	--	--
3.	Methanol-Carbon disulfide	11.95	12.15	8.80	8.80	2.75	2.70	--	--
4.	Methanol-diethyl ether	10.70	--	7.30	7.27	--	--	--	--

IV. CONCLUSIONS

Variations in the values of individual parameters are a result of interference by parameters other than the one to be evaluated. Value for δ_h are all unreasonable. The δ_d for diethyl ether-methanol and hexane-methanol are locked into their value by virtue of the δ_d of the nonsolvents. The value of δ_d calculated from hexane-carbon disulfide is high probably because the δ_p and δ_h are included. However, values of δ_d and δ_h for methanol and carbon disulfide appear to be somewhat realistic. In any case, values for δ_o in the neighborhood of 11.0 and δ_{d3} around 8.8 and δ_{p3} around 3.6 seem realistic but value of δ_{h3} above 5.0 seems out of line.

The use of more carefully selected nonsolvents should help to iron out the variations.

V. ACKNOWLEDGEMENTS

This investigation was supported in part by funds supplied by an ACS-PRF grant.

VI. REFERENCES

1. Suh, K.W. and Clarke, D.H., Polym. Sci., 5, 1671 (1967).

2. Hansen, C.M., J. Paint Technol. 39, No. 505, 104 (1967).

3. Hansen, C.M., Skaarup, K., J. Paint Technol., 39, No. 511, 511 (1967).

4. Hansen, C.M., J. Paint Technol., 39, No. 511, 505 (1967).

5. Hildebrand, J.H., Scott, R.L.,"The Solubility of Non-electrolytes,"Dover Publications, Inc., New York, 1964, page 129.

6. Hildebrand, J.H., ibid., 370 (1964).

7. Scott, R.L., J. Chem. Phys., 17, 268 (1949).

8. Hansen, C.M. and Beerbower, A., "Kirk-Othmer Encyclopedia of Chemical Technology", Suppl. Vol., 2nd Ed., 889 (1971).

SOLUBILITY OF STYRENE-ACRYLONITRILE COPOLYMERS

R. F. Blanks, B. N. Shah
Michigan State University

I. INTRODUCTION

Thermodynamic phase equilibrium data are needed in many polymer engineering situations. Some of these include: Correlations of efficiency of polymer devolatilization equipment; Analysis of pressure control systems in polymerization reactors; Description of kinetics and phase equilibria in two-phase polymer systems; Prediction of solubility relationships in coatings and plasticizer system formulations; Estimation of swelling resistance in materials selection. Thermodynamics also plays an important role in influencing the size and shape of macromolecules in solution. This, as well as the kinetics of entanglement formation, influence transport phenomena, viscosities, and diffusivities, in macromolecular solutions.

A first approximation to the description of polymer solution thermodynamics is the Flory-Huggins equation, Equation 1, which represents the molar Gibbs energy of mixing polymeric solute with monomeric solvent.

$$\Delta G_m = RT[n_1 \ln \phi_1 + n_2 \ln \phi_2 + n_1 \chi \phi_2] \tag{1}$$

In Equation 1, n_1 and n_2 represent the number of moles of solvent and polymer respectively and the ϕ_i's are volume fractions. This equation contains a polymer-solvent interaction parameter, χ, which is characteristic of the intermolecular forces between the molecules in solution, and which has a magnitude depending upon both entropy and enthalpy effects. More recent and more complex theories of polymer solutions often result in expressions containing concentration dependent free energy functions [1,2]. At low concentrations the interaction parameter, or free energy function, should depend upon excess entropy of mixing and upon heat of mixing polymer with solvent. In this case, the parameter χ may be considered as a residual chemical potential. The heat of mixing may be estimated from three dimensional solubility parameter concepts [3]. Excess entropy of mixing and free volume effects are not yet quantitatively predictable in the absence of thermodynamic data for a mixture. Patterson [4] has shown the similarity of the solubility parameter theory and the corresponding states theory.

From Equation 1 the activity of the solvent in a binary polymer-solvent mixture is

$$\ell n\ a_1 = \ell n\ \phi_1 + (1 - v_1/v_2)\ \phi_2 + \chi\ \phi_2^2 \tag{2}$$

where v_1/v_2 is the ratio of the molar volumes of the solvent and polymer, or approximately the reciprocal of the degree of polymerization of the polymer. The only adjustable parameter in this treatment of polymer solutions is the interaction parameter, χ. If all solution non-idealities are lumped into this parameter Equation 2 may be written as [5]

$$\ell n\ a_1 = (\ell n\ a_1)_{combinatorial} + (\ell n\ a_1)_{noncombinatorial}$$

$$= [\ell n\ \phi_1 + (1 - v_1/v_2)\ \phi_2] + \chi\ \phi_2^2 \tag{3}$$

$$\chi = (\ell n\ a_1)_{noncombinatorial} / \phi_2^2 \tag{4}$$

$$\chi = \chi\ (\phi_2, T) \tag{5}$$

Bonner and Prausnitz [6] have shown how the Prigogine-Flory theory [1,2] may be used to perform vapor liquid equilibrium calculations for concentrated polymer solutions. In order to use the free volume theory three molecular parameters are needed for each pure component, along with one mixture parameter in the case of a binary system. The pure component parameters may be obtained from density, pressure, temperature data [7], or estimated by use of the correlations of Bondi[8].

In essence the Prigogine-Flory theory develops from a corresponding states point of view, an equation of state for the pure components and the mixture. The activity coefficient of one component in the mixture is then formulated as a sum of contributions from; (1) an athermal, combinatorial entropy of mixing, (2) a free-volume term, and (3) an intermolecular force term. In principle the parameters from this corresponding-states theory are concentration independent, which is in itself an improvement over the older Flory-Huggins lattice theory [9].

As shown by Bonner and Prausnitz correlations of mixture data are very sensitive to the binary mixture parameter contained in the theoretical development. Furthermore, if the Prigogine-Flory theory is used for relatively expanded liquids the characteristic molecular parameters for pure substances may loose some of their physical meaning. This newer work, based upon the statistical-mechanical theory of Prigogine [10] appears to be helpful in describing the combined influence of free volume and intermolecular forces upon excess free energy of mixing in the case of polymer solutions. However, data of the required form, and analysis, for copolymer solutions are not readily available.

This paper reports measurements of the interaction parameter, χ, Equation (1). Interpretation of the results are based upon the three dimensional solubility parameter concept. In dilute solutions the thermodynamic interaction parameter may be measured by light scattering techniques, and may be estimated from intrinsic viscosity measurements. The solubility of a polymer in a solvent may then be expressed in terms of intrinsic viscosity, or the second virial coefficient from light scattering. A "good" solvent for a polymer is one in which polymer segments prefer contacts with the solvent molecules and the polymer expands or swells in solution as opposed to a "poor" solvent in which the polymer segments prefer contact with their own kind and thereby the polymer molecule tends to coil-up and minimize the volume of solution it occupies. A good solvent for a polymer is one in which the residual chemical potential is a minimum. In such systems the second virial coefficient from light scattering, A_2, and the intrinsic viscosity, $[\eta]$, have maximum values. Also the swelling ratio, α, which is the ratio of the dimensions of the swollen polymer chain in a good solvent, to the dimensions of the same polymer chain in a poor solvent at incipient precipitation, is a maximum in the "best" solvent.

In this work dilute solution solubility data were obtained for styrene-acrylonitrile copolymers, SAN copolymers, of four different acrylonitrile contents in four solvents of different polarity.

II. EXPERIMENTAL

Four samples of SAN copolymer and two samples of polystryrene were prepared, each having different molecular weight and different ACN content. These samples were prepared in bulk at 60°C, utilizing α-α' Azobisisobutyronitrile as initiator. The monomers and initiator were purified using appropriate procedures. Reactions were carried out, with agitation, to low conversions in order to minimize copolymer composition drift during copolymerization. The uniformity of the chemical composition of the copolymer samples was verified by light scattering experiments. The copolymer samples were analyzed for nitrogen content, and thus ACN content, and for molecular weight and molecular weight distribution by gel permeation chromatography. Table 1 describes the polymer and copolymer samples prepared for this investigation.

TABLE 1
Description of Polymer and Copolymer Samples

Polymer	Composition, ST		M_w 10^{-5}		$M_n 10^{-5}$	M_w/M_n
	Weight %	Mole %	GPC	L.S.*	GPC	GPC
PS-1	100	100	1.85	1.91	1.03	1.79
PS-2	100	100	5.01	5.04	2.41	2.08
SAN C-1	85.8	74	2.75	2.90	1.41	1.95
SAN C-2	76.0	62	2.03	1.80	1.20	1.69
SAN C-2'	77.0	62	6.34	6.66	3.39	1.87
SAN C-3	62.0	48	3.32	3.32	2.05	1.62

*Gel Permeation Chromatography, Dow Chemical Laboratories; L. S., Light Scattering, Michigan State University Laboratories.

The four solvents used in this work were benzene, dioxane, methyl ethyl ketone, and dimethyl formamide. The thermodynamic quality of each solvent for each polymer, or copolymer, was determined from light scattering and intrinsic viscosity measurements. Light scattering experiments were made using a Brice-Phoenix light scattering photometer of the series 2000, located in a low-humidity, constant-temperature room. Twelve scattering angles and four polymer solution concentrations were used for each system investigated in these experiments. Specific refractive index increments for each polymer-solvent pair were determined using a Brice-Phoenix Differential Refractometer Model BP-2000-V. Second virial coefficients were obtained from Zimm plots such as the one shown in Figure 1.

Intrinsic viscosities were measured in Cannon-Ubbelohde suspended level U-tube capillary viscometers of the four bulb type, which allows obtaining readings at four different shear rates per sample. Temperature of measurement was held constant by clamping the viscometers vertically in an insulated water bath equipped with a precision thermostat. Before introducing the solutions into the viscometers, they were clarified by pressure filtration through ultrafine sintered glass filters. Results are shown in Table 2.

III. DISCUSSION

In a solution a polymer molecule is often modeled as a long-chain necklace, or as a spherical cloud of polymer segments. The end-to-end distance between ends of the polymer chain, or the radius of the spherical cloud of segments, tend toward maximum values in good solvents and are a minimum in

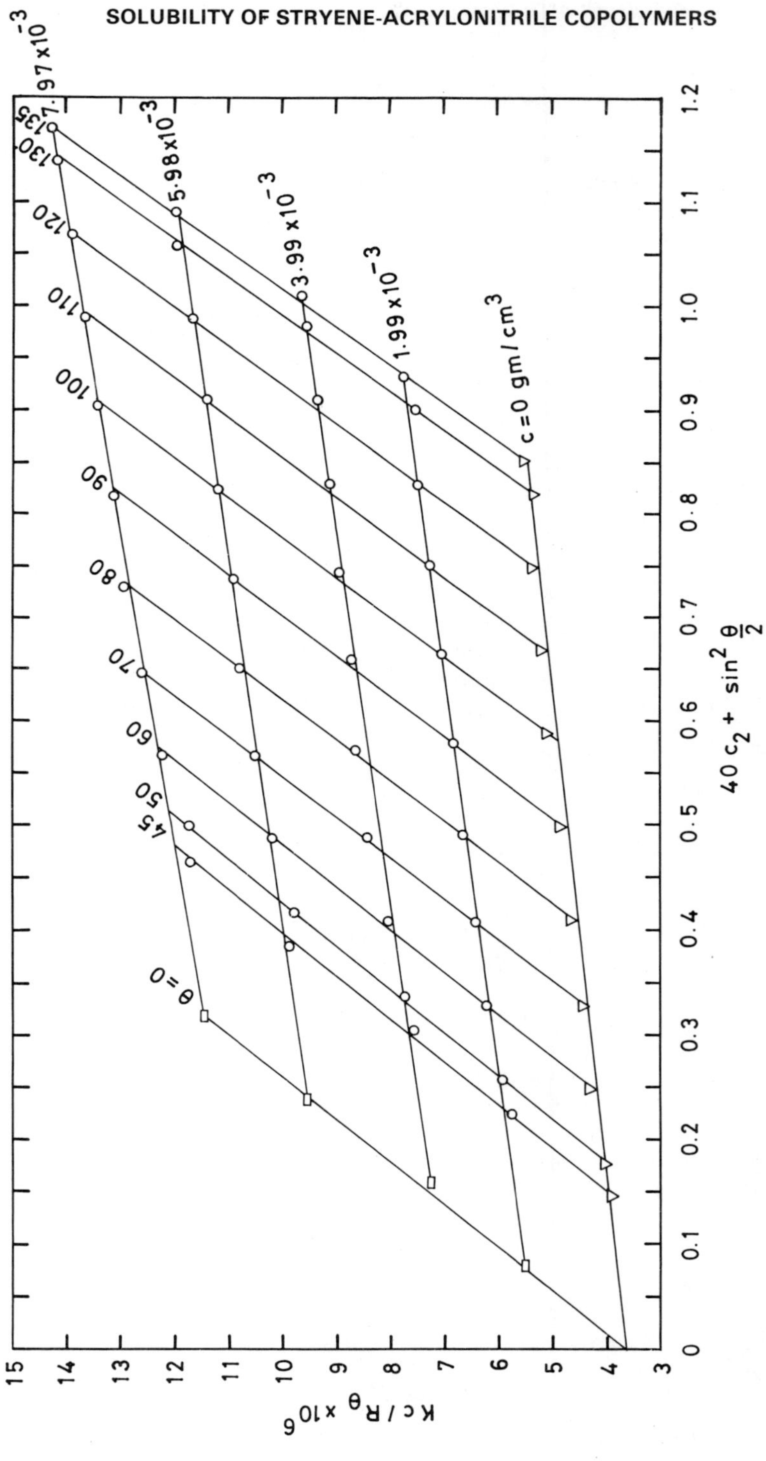

Fig. 1. Zimm Plot for SAN C-1 in DMF at 25°C.

TABLE 2
Thermodynamic Parameters

Polymer	Solvent	$A_2 \times 10^4$ mole cc/gm^2	χ^a	$[\eta]$ dl/gm	α^b
PS-1	Benzene	5.64	0.426	0.782	1.278
	Dioxane	4.83	0.443	0.622	1.184
	MEK	1.73	0.481	0.433	1.050
PS-2	Benzene	5.71	0.424	1.619	1.380
	Dioxane	--	--	1.503	1.346
	MEK	--	--	0.785	1.084
SAN C-1	Dioxane	6.82	0.418	1.007	1.150
	Benzene	5.36	0.433	0.988	1.143
	DMF	4.80	0.450	0.867	1.094
	MEK	4.00	0.453	0.783	1.058
SAN C-2	DMF	6.92	0.426	0.862	1.185
	Dioxane	6.10	0.426	0.816	1.164
	MEK	5.70	0.429	0.742	1.128
	Benzene	-0.66	0.507	0.446	0.952
SAN C-2'	DMF	6.76	0.420	2.110	1.287
	Dioxane	6.00	0.426	1.987	1.261
	MEK	5.64	0.430	1.776	1.215
	Benzene	-0.68	0.507	0.725	0.901
SAN C-3	DMF	10.10	0.382	1.653	1.218
	MEK	5.66	0.428	1.219	1.107
PAN	DMF	19.10[b]			

a. The interaction parameter, χ, was obtained from A_2 and α

$$A_2 = ((\tfrac{1}{2}-\chi)/v_1\rho_2{}^2)\left(1 - \frac{x}{2!\,2^{3/2}}\right)$$

where $x = 2\ (\alpha^2 - 1)$.

b. Expansion factors were calculated, based upon unperturbed dimensions of SAN copolymers using an interpolation of the data of Shimura [11].

a poor solvent. The condition where the polymer molecule's dimensions in solution are a minimum is termed a theta condition. The dimensions of a polymer molecule in solution are influenced by both short-range and long-range interactions.

The short-range interactions result from side group steric effects and bond angle restrictions. These short-range interactions are reflected in a parameter σ, the steric factor or stiffness parameter, which is a measure of the hindrance to internal rotation about the carbon-carbon single bonds of the backbone, and which generally increases with the molar volume of the substituent groups on the main chain of a linear polymer. The steric factor is defined as the ratio of the root-mean-square end-to-end distance of the polymer chain in the θ state to the hypothetical root-mean-square end-to-end distance of a chain in which the internal rotation about the carbon-carbon bond of the main chain is completely free. Shimura [11] has calculated values of σ from experimentally determined values of θ condition parameters and found that the dimensions of random copolymers of styrene and acrylonitrile in θ solvent are not significantly influenced by the copolymer composition. Values of σ were 2.22 for polystrene, 2.20 for polyacrylonitrile, and 2.25 to 2.31 for SAN copolymers.

In the systems investigated in this work the dimensions of the polymer molecule in solution are determined by long-range thermodynamic interactions between polymer and solvent. These are reflected in the values of the expansion factor α, listed in Table 2. The expansion factor α represents the ratio of the polymer molecules dimensions in the solvent system of interest to the value in a theta condition. Also in Table 2 it may be observed that larger α values are associated with larger values of the second virial coefficient, A_2, and that for a particular polymer dissolved in a particular solvent the lower the value of the free energy parameter, χ, the better the solvent for the polymer. Table 2 lists the polymer solvent systems in descending quality of solvent for each polymer.

The second virial coefficient may be considered as a measure of long-range interactions between polymer and solvent. The dependence of A_2 on molecular weight is negligibly small for the polymers and copolymers over the range of molecular weights considered in this work [11]. Therefore, the variation of A_2 with polymer composition in different solvents is an indication of the variation of solvent quality with copolymer composition. The higher the value of A_2 for a polymer in a solvent, the better is that solvent for that polymer. As the solvents for a polymer become poorer, the values of A_2 in those solvents become lower and finally at the θ condition, A_2 becomes zero. Figure 2 shows the plots of A_2 of the polymers in different solvents against copolymer composition. The

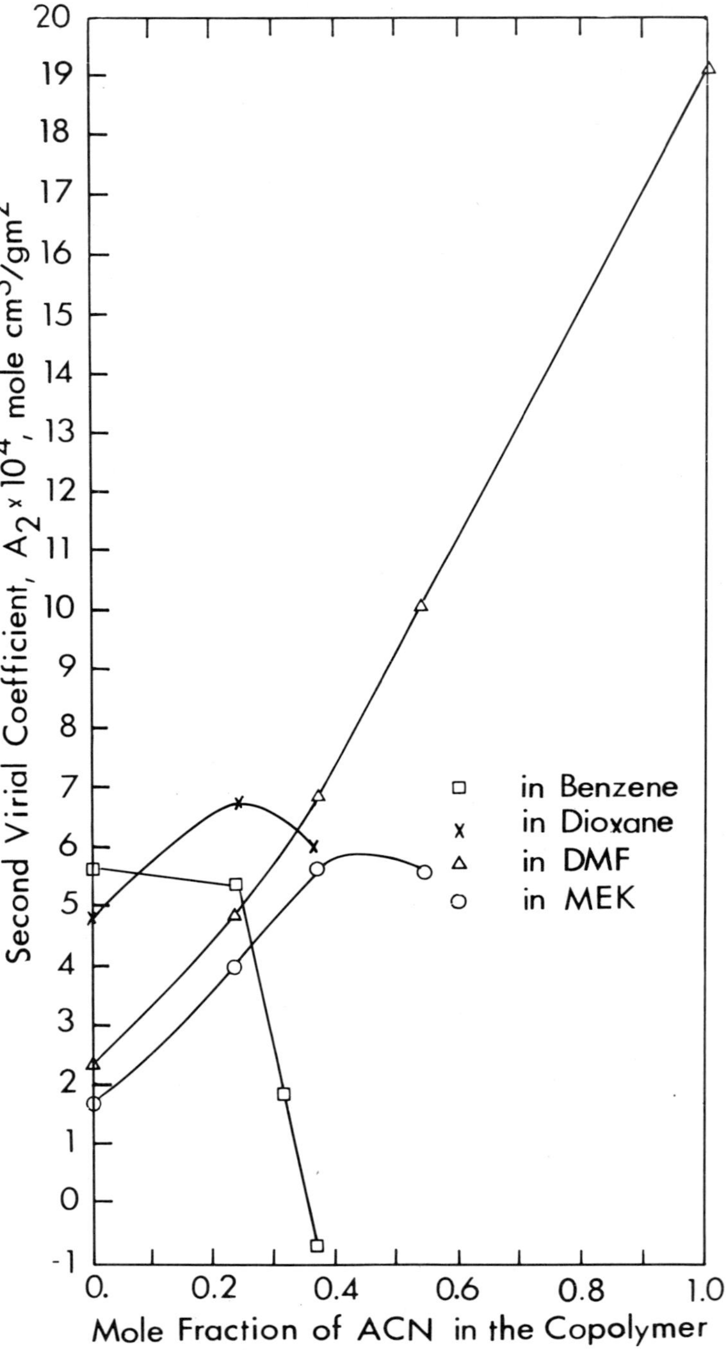

Fig. 2. Thermodynamic Quality of Solvents for SAN Copolymer.

points are joined to emphasize the trend in behavior.

For copolymers in MEK, solvent quality is a maximum in the neighborhood of 0.5 mole fraction ACN in the copolymer. This was also observed by Shimura [11]. This maximum at approximately equimolar composition occurs where there exists the greatest tendency of alternation of styrene and acrylonitrile groups in the copolymer. The behavior in dioxane shows a similar trend, with the maximum in solvent quality occurring at about 0.25 mole fraction of ACN. Both of these solvents are nonsolvents for polyacrylonitrile. With the ACN content higher than that in SAN C-3, MEK would be a nonsolvent, however this limiting concentration is not predicted quantitatively. Dioxane does not dissolve the copolymer SAN C-3 which has a 0.52 mole fraction ACN content. Both dioxane and MEK become better solvents for copolymers with increased ACN content up to a certain ACN content, and then decrease in quality. This may be due to the polarity of MEK and localized polarity of dioxane.

Benzene is a non-solvent for polyacrylonitrile. This non-polar solvent is a poorer solvent than MEK or dioxane for copolymers with ACN content higher than 0.24 mole fraction. This is evident from the deep plunge in the value of A_2 for SAN C-2. Here the second virial coefficient is negative, indicating that the copolymer is in an extremely poor solvent environment. The copolymer molecules are tightly coiled, or clustered together. In an environment of benzene the chains prefer polymer-polymer contacts rather than polymer-solvent contacts. The tightly coiled nature of the chains is also indicated by the α value, expansion factor in benzene, which is less than unity.

Dimethyl formamide is a good solvent for polyacrylonitrile and a relatively poor one for polystyrene. Values of the second virial coefficient for this system increase continuously as ACN content increases. This may be an indication of strong localized interactions between DMF and ACN moities in solution.

Solvent quality improves as the free energy paramater χ decreases. Decreases in the parameter χ should be accompanied by decreases in the interchange energy density, A_{12}. The value of A_{12} is characterized by the difference between polymer-polymer plus solvent-solvent interactions and polymer-solvent interactions. These interactions depend upon non-polar dispersion forces, dipole-dipole interactions between polar molecules, and specific interactions between molecules such as hydrogen bonding or charge transfer complex formation [12]. The interchange energy density may therefore be formulated in terms of solubility parameters for the polymer and solvent.

A large number of solvents have been considered by Hansen and Beerbower [3] and non-polar, polar, and specific interaction solubility parameter values are tabulated in the Encyclopedia of Chemical Technology. These are given the symbols of δ_d, δ_p, and δ_H. Using the concept of molecular homomorph and the molar-attraction constants of Small [13] or Hoy [14] the dispersion, polar, and hydrogen bonding solubility parameters of polymers may be estimated [12]. Using ethyl benzene and propionitrile as homomorphs for the polystyrene and polyacrylonitrile, three-dimensional solubility parameters were calculated for the polymers and copolymers used in this work, based upon the assumption that copolymer solubility parameters are simple volume fraction averages of homopolymer solubility parameters [15]. Solubility parameters for the systems investigated in this work are shown in Table 3.

TABLE 3
Three dimensional Solubility Parameters

Substance	δ_d	δ_p	δ_H
Benzene	9.0	0	1.0
Dioxane	9.3	0.9	3.6
MEK	7.8	4.4	2.5
DMF	8.5	6.7	5.5
Polystyrene	9.1	0.3	0.7
Polyacrylonitrile	9.0	8.4	3.1
SAN C-1	9.1	1.4	1.0
SAN C-2	9.1	2.1	1.2
SAN C-3	9.1	3.0	1.5

Values of the enthalpy contribution to χ may then be obtained as

$$\chi_H = v_1/RT[(\delta_{d1}-\delta_{d2})^2 + 0.25((\delta_{p1}-\delta_{p2})^2+(\delta_{H1}-\delta_{H2})^2)] \quad (6)$$

where v_1 is solvent molar volume. Such calculations for χ_H do predict the trends indicated in Figure 2 for various copolymers in a single solvent. Ranking of solvents for a single copolymer, or quantitative predictions of χ have not yet been achieved.

Table 4 compares the calculated enthalpy contribution to the interaction parameter Equation 6, with the experimental interaction parameter values. Previous estimates of the interaction parameter χ had often been obtained by adding the enthalpy portion, χ_H, to a constant entropy portion $\chi_S = 0.34$ [12]. Patterson has indicated some justification for this empiricism and indicates how an entropy term χ_S may be used to reconcile solubility parameter theory with the newer

TABLE 4
Interaction Parameter Calculations

System	χ Experimental	χ_H Calculated
Polystyrene		
Benzene	0.43	~ 0
Dioxane	0.44	0.3
MEK	0.48	1.0
DMF	--	2.1
SAN C-1		
Dioxane	0.42	0.3
Benzene	0.43	0.1
DMF	0.45	1.6
MEK	0.45	0.7
SAN C-2		
DMF	0.43	1.3
Dioxane	0.43	0.3
MEK	0.43	0.5
Benzene	0.51	0.2
SAN C-3		
DMF	0.38	1.0
MEK	0.43	0.4
Dioxane	--	0.3
Benzene	--	0.4
PAN		
DMF	--	0.3
MEK	--	0.8
Dioxane	--	2.1
Benzene	--	2.8

corresponding states concepts [4]. The calculations in Table 4 show that a constant value for the entropy contributions is not justified for these polar systems. Free volume effects must be accounted for in a more fundamental manner. Furthermore the calculations suggest that a specific interaction parameter including both the Lewis acid and the Lewis base tendency for interaction should be used to characterize intermolecular forces. In other words, the relative positions of the electrons and protons in portions of molecules depend upon the molecules environment. Thus DMF and MEK (enol form), may act as proton donors in respect to SAN copolymer more strongly than they do with respect to molecules of like kind in their own pure solvent state. This means that single values of a hydrogen bonding parameter based upon the energy of vaporization of pure solvent may not adequately describe the hydrogen bonding tencency of the solvent in a polymer solution. For this reason DMF and MEK are predicted to be non-solvents for SAN copolymers but in fact are good solvents, at least in certain copolymer composition ranges.

Additional experimental data for copolymer solutions, particularly concentration and temperature effects, and analysis of the thermodynamics of copolymer solutions based upon the Prigogine-Flory theory of corresponding states appears to be a desirable continuation of this work.

IV. REFERENCES

1. Patterson, D., Rubber Chem. and Technol. 40, 1 (1967).
2. Flory, P. J., Disc. Faraday Soc. 49, 7 (1970).
3. Beerbower, A., and Hansen, C. M., in "Kirk-Othmer Encyclopedia of Chemical Technology," 2nd ed., Supplement Volume p. 889 (1971).
4. Biros, J., Zeman, L., and Patterson, D., Macromol. 4, 30 (1971).
5. Patterson, D., *et al.*, Macromol. 4, 356 (1971).
6. Bonner, D. C., and Prausnitz, J. M., AIChE Journal, 19 943 (1973).
7. Eichinger, B. E., *et al.*, Coatings and Plastics Preprints, 35, 301 (1975).
8. Bondi, A., "Physical Properties of Molecular Crystals, Liquids, and Glasses," p. 214. Wiley, New York, 1968.
9. Flory, P. J., J. Chem. Phys. 10, 51 (1942).
10. Prigogine, I., "The Molecular Theory of Solutions." North-Holland, Amsterdam, 1957.
11. Shimura, Y., J. Polym Sci., Part A-2 4, 423 (1966).
12. Blanks, R. F., and Prausnitz, J. M., I&EC Fund. 3, 1 (1964).
13. Small, P. A., J. Appl. Chem. 3, 71 (1953).
14. Hoy, J. Paint Technology, 42, 76 (1970).
15. Krause, S., J. Marcomol. Sci-Revs. Macromol. Chem. C7(2) 251 (1972).

ACKNOWLEDGMENTS

The authors wish to acknowledge the donors of the Petroleum Research Fund, administered by the American Chemical Society for partial support of this research. Also thanks to Jack Kinsinger and Bill Deal of Michigan State University and to the Dow Chemical Company and Turner Alfrey, for support and discussion.

SOLUBILITY OF HYDROGEN IN POLYETHYLENES OF DIFFERENT MORPHOLOGICAL STRUCTURE. II. SINGLE CRYSTALS, EXTENDED CHAIN CRYSTALS AND FIBERS

Mark E. Land and Malcolm Dole
Baylor University

Using a new quenching method (1) Kubo and Dole (2) were able to demonstrate a greatly enhanced solubility of the permanent gases, hydrogen and helium, in single crystal mats of polyethylene (PE) as compared to bulk linear PE. Calculated on the assumption that the gases dissolved only in the amorphous phase, the solubility per unit of amorphous phase was 10 fold greater for hydrogen and 18 fold greater for helium at 25° in the single crystal mats than in bulk PE. Even if the calculations were not restricted to the amorphous phase, the solubility per gram of sample was 5 fold greater in the case of hydrogen and 9 fold greater for helium in the single crystalline mats as compared to the bulk PE.

Ambient hydrogen gas also has a marked catalytic effect (3) on the first order decay of the residual alkyl radicals in irradiated PE, and because of the fact that the catalytic effect was about 10 fold greater in the case of the single crystalline mat as compared to bulk linear PE, it was believed that the greatly enhanced solubility of hydrogen in the single crystalline mats was a true solubility. However, we have recently studied the hydrogen solubility in freeze-dried single crystals of different fold periods. Much to our surprise no hydrogen solubility could be detected in these rather fluffy samples. Yet the hydrogen still catalyzed the alkyl radical decay and to the same extent as in the case of the single crystalline mats (4). This observation led us to extend our solubility measurements to single crystal samples prepared in different ways and having different fold periods. Further work has also been done on extended chain samples and on polyethylene fibers.

I. EXPERIMENTAL SECTION

The technique was the same as that already described (1). The chief difficulty was the occurrence of temporary leaks in the solubility cell during the heating from 77K to 25° which resulted from the different thermal coefficients of expansion of the stainless steel valves and the copper tubing at the point where they were joined together by stainless steel ferrules of the Swagelock fittings. The difficulty was overcome by having a commercial welding company silver-solder the ferrule to the tubing.

Materials studied included freeze-dried samples (5) of single crystalline PE formed by crystallization from a 0.2 wt % solution of Marlex-6002 PE in xylene at 70, 80 and 90°. After crystallization the slurry of crystals and solution was rapidly frozen in liquid nitrogen and the solvent molecules mechanically pumped off. Hot-filtered PE mats of single crystals were prepared in this laboratory and the crystals were separated from the solution by filtration at the temperature of crystallization. The latter samples were dried for 24 hrs in a vacuum oven at 60°. Some PE fibers of 976 denier monofilaments made of low density PE were obtained from Amtech, Inc. A film was made of this material by melting some of the fibers between glass plates in a vacuum oven at 140°. X-ray diffraction patterns of the film indicated that some fiber structure still persisted in the annealed film. The extended chain sample was the same as the sample P-70 used by Kubo and Dole (2). The properties of the samples used are listed in Table I where α_w is the weight fraction of the amorphous phase.

TABLE I
Properties of Polyethylene Samples Studied

Type of Sample	Density ρ at 25° g cm^{-3}	α_w
Film	0.960	0.242
Fibers	0.922	0.494
Film from Melted Fibers	0.920	0.506
Extended Chain	0.986	0.078

Single Crystal Samples

Designation	Temp. of Crystallization	Fold Period, ℓ A-unit	Treatment	Density ρ at 25°	α_w
SC-70	70	112	Hot-filtered	0.971	0.170
SC-90	90	150	Hot-filtered	0.976	0.141
SC-70-F	70	112	Freeze-dried	0.971	0.170

II. DATA AND DISCUSSION

Because of a parallel study in this laboratory of the kinetics of free radical decay in single crystal samples of PE having different fold periods and of the catalytic effect of hydrogen gas in promoting the decay, we were interested in determining the solubility of hydrogen in these samples. The first samples studied were the freeze-dried samples, and much to our surprise no solubility could be detected although three attempts were made to measure the solubility, two at 25° and one at 35°, all at 1 atm. pressure of hydrogen. The freeze-dried samples were rather fluffy, apparently with little contact between crystalline lamellae. It appears, therefore, that for the single crystalline samples to exhibit hydrogen solubility, there must be at least two lamellae packed rather tightly together with an amorphous region between them in which hydrogen can dissolve and will be retained during the evacuation of the sample at 77K, the procedure used (1) to remove the ambient hydrogen gas. The influence of the packing of the lamellae on formation of PE gels by radiation cross-linking is already well known [for a review see Salovey (6).]

The observation of no detectable hydrogen solubility in the freeze-dried samples demonstrates that there are few if any sorption sites on the surface of these crystals. This means that the chain folds on the surface must be, for the most part, tight folds with no cavities sufficiently big to accommodate hydrogen molecules.

Data are given in Table II for hydrogen and deuterium solubilities in mats of hot-filtered single crystalline samples of linear PE and for the other samples studied. σ_a values are in units of p.p.m. by weight and were calculated from the measured solubility, σ_c, by dividing the latter by the weight fraction of amorphous content, α_w. The σ_a values represent, therefore, the solubility calculated assuming that all of the hydrogen is dissolved in the amorphous phase. The ratio of σ_a for deuterium to $2\sigma_a$ for hydrogen is nearly unity in all of the samples although, because of the difference in the heats of solution between H_2 and D_2, the ratio changes slightly with temperature. As the internuclear distances in H_2 and D_2 are practically identical according to data in the tables of Herzberg (7), 0.7417 A° for H_2 and 0.7416 for D_2, we can assume that the volumes occupied by the H_2 and D_2 molecules are the same.

Making use in part of the theory of Frisch and Rogers (8), Ziegel and Eirich (9) have calculated the cavity free

TABLE II
Solubility Data at 1 Atm. Pressure

Sample	Temp.	σ_c		σ_a	$D_2/2H_2$
		g/g PE x 10^6			
		H_2	H_2	D_2	
SC-70	0	0.796	4.68	9.99	1.07
	25	.796	4.68	9.69	1.04
	35	.808	4.75	9.56	1.01
	45	.804	4.73	9.72	1.03
SC-90	25	1.99	14.1	28.2	1.00
Extended Chain	0	0.544	6.98	14.9	1.07
	25	.552	7.08	14.1	1.00
	35	.548	7.02		
	45	.547	7.01	13.8	0.98
Fiber	0	1.08	2.18	4.57	1.05
	25	1.26	2.56	4.92	.96
	35	1.33	2.70		
	45	1.37	2.78	5.10	.92
Film from Fiber	0	1.30	2.56	5.17	1.01
	25	1.51	2.98	6.08	1.02
	35	1.64	3.24	6.30	.97
	45	1.76	3.48	6.65	.96

TABLE II (CONTINUED)
Solubility Data at 1 Atm. Pressure

Sample	Temp.	σ_c	σ_a		$D_2/2H_2$
		g/g PE x 10^6			
		H_2	H_2	D_2	
Marlex-6002 Film	0	0.880	3.68	7.59	1.03
	25	.927	3.88	8.08	1.04
	35	.946	3.96	8.20	1.04
	45	.961	4.02	8.28	1.03

volume, v_f, from the equation

$$v_f = \left[\frac{18.8}{(-\delta\Delta H_s)^{\frac{1}{2}}} + d\right]^3 \quad (1)$$

in which $\delta\Delta H_s$, which is always negative, represents $\Delta H_s(D_2) - \Delta H_s(H_2)$ and d is the diameter of the probe molecule. It is assumed that d for H_2 is equal to that for D_2 and in eq 1 is expressed in Angstrom units, or 2.34 A. Eq 1 is derived from the quantum mechanical equation for a particle in a box assuming that the particle is in its ground translational state (valid for a steep potential energy well.) A shape factor of three valid for a cubical cavity is also used. The constant factor 18.8 is calculated for ΔH_s in units of calories per mole. In Table III we have collected the heats of solution of H_2 and D_2 and the cavity free volumes calculated using eq 1. The heats of solution were determined by a least squares evaluation of the slope of a $\ell n\ \sigma_a$ versus 1/T plot in the usual manner. An example of such plots is given in Figure 1 for H_2 and D_2 solubilities in the fiber and film from fiber samples.

Note that all of the heats of solution of H_2 are endothermic; that is, heat is absorbed during the solution process and the solubility increases with the temperature. Also the heat is considerably more endothermic for the hydrogen solubility in the fiber and the film from fiber samples.

TABLE III
Heats of Solution and Calculated Cavity Free Volumes

Sample	ΔH_s cal mole^{-1} H_2	D_2	v_f A^3	σ_a at 25° for H_2
SC-70	55.0	-136.0	50.6	4.68
SC-90				16.0
Marlex-6002 Film	341.8	313.7	204	3.88
Fiber	951.2	422.8	31.5	2.56
Film from Fiber	1181	957.7	46.7	2.98
Extended Chain	40.0	-308.0	37.6	7.08

The PE used in making these samples was a low density, branched type of polyethylene. There is a rough correlation between the heat of solution and the solubility, better in the case of D_2 than H_2. The more endothermic the heat solution the less the solubility, see Figure 2 where data are plotted for the D_2 solubility and heat of solution values.

With the exception of the value of v_f for the Marlex-6002 film sample, the cavity volumes are all in the neighborhood of 30-50 A^3; but there does not seem to be any correlation between the solubility and cavity size. In fact, if the cavity volume is divided into the total free volume of the Marlex-6002 film, for example, we calculate many more cavities than there are hydrogen molecules dissolved at one atmosphere. Thus, taking the total free volume to be the difference in specific volume of the sample and that of the pure PE crystal, or 0.041 cm^3 g^{-1} and dividing this by the volume of one cavity in the case of the Marlex-6002 film or 2.04×10^{22} cm^3 we find that there are 2.0×10^{20} cavities. At one atm. pressure of H_2, the number of molecules of H_2 dissolved per gram of sample is
$(3.88 \times 10^{-6})(0.242)(6.02 \times 10^{23})/2.018$ or 2.8×10^{17}

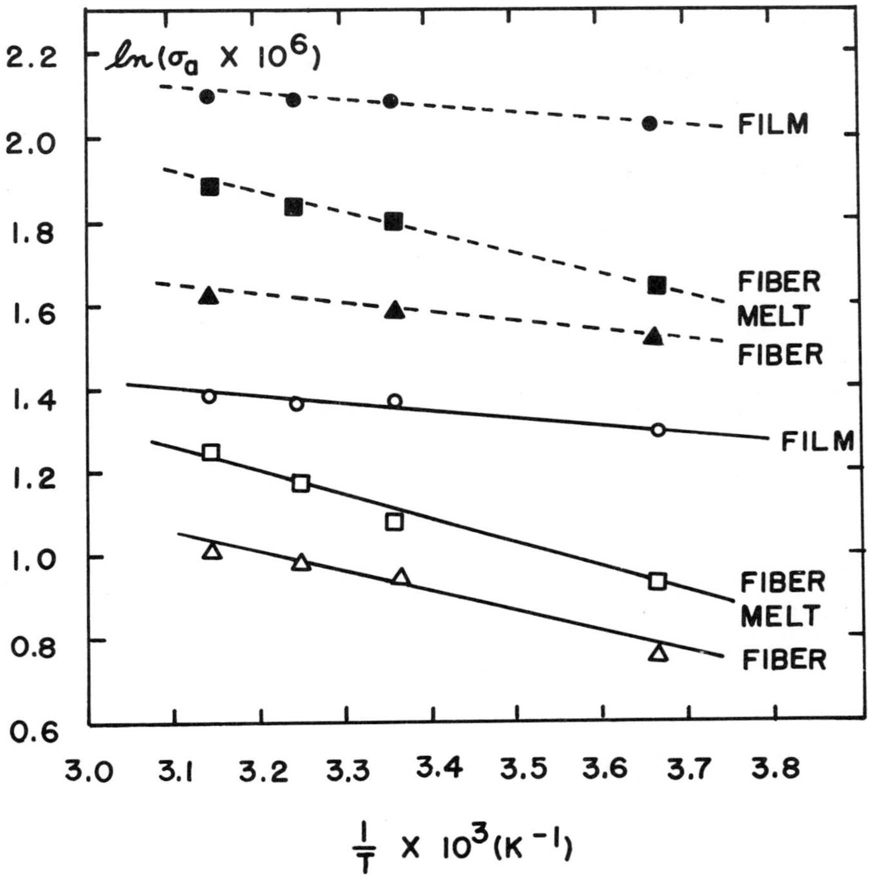

Fig. 1. Natural logarithm of hydrogen and deuterium solubilities in the fiber and film from fiber samples plotted as a function of reciprocal temperature. Dotted lines represent deuterium solubilities.

molecules g^{-1}. Thus there are about 1000 more cavities than hydrogen molecules dissolved. Of course, the cavity volumes are undoubtedly not uniform so that there may be many cavities too small to accommodate a hydrogen molecule and many larger than the above estimated cavity volume. The ratio of total free volume to cavity volume would be even greater in the case of the other samples studied.

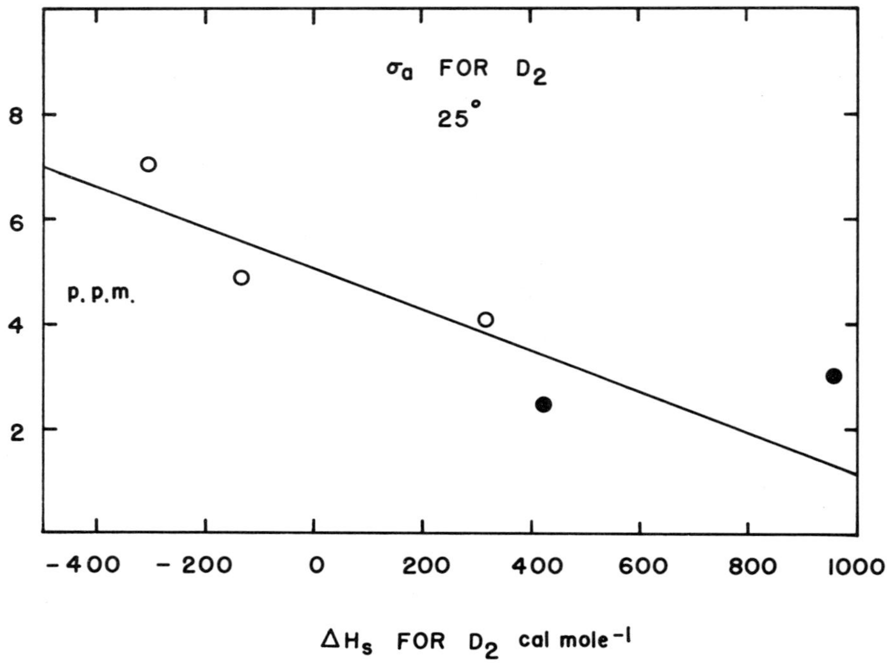

Fig. 2. Solubility of D_2 gas in parts per million by weight at one atm. pressure and 25°, in different types of PE as a function of the heat of solution.

Another point of view is to consider that each end of a polymer chain serves as a sorption site. We have calculated the ratio of the solubility of hydrogen in terms of moles per gram to the moles of ends per gram for the n-paraffinic hydrocarbons listed in our previous publication (2). For the n-paraffins the ratio is practically constant from C_6H_{14} to $C_{16}H_{34}$ at about 3 to 4 x 10^{-4}, but rises to 190 x 10^{-4} in the case of PE. The much larger ratio for PE could be either the result of more sorption sites along the chain or of a greater accessibility of the hydrogen molecules to the chain ends, but it is probably the former because the hydrogen solubility per unit of amorphous phase in the PE is very close to that predicted (2) by extrapolation of the solubility data of the low molecular weight n-paraffins to infinite molecular weight.

Turning now to the hydrogen solubility in the fiber sample, we see, Table II, that the solubility per unit of amorphous phase is the lowest of all of the samples studied. As the fiber was made from branched PE we melted some of the fiber and pressed it into film form in order to determine the effect of fiber structure on the hydrogen solubility. The σ_a value in the film from the fibers was also low; however, the actual solubility, σ_c, was greatest in the fiber and fiber film (except for the hot-filtered SC-90 sample.) The low values calculated for σ_a were the result of high values of α_w in the fiber and fiber film. In making this calculation we have assumed that the weight fraction of the amorphous phase is correctly calculated from the density values. In other words we assume that the density of the amorphous phase is the same in all of the different samples studied. Fischer et al (10) have estimated densities of the crystalline regions in drawn films before and after annealing from low angle X-ray scattering data. They concluded that calculation of the fraction of crystallinity from density values is meaningless because of the fact that the effective density of the crystalline region, ρ_c, in the drawn samples is less than that calculated from the unit cell dimensions of the ideal crystal. On annealing, the density difference, $\Delta\rho$, ($\Delta\rho = \rho_c - \rho_a$) approaches that of the ideal value 0.142 (at 30°). If $\rho_c - \rho_a$ for the fiber is smaller than the ideal value, then on calculating the amorphous volume fraction α_v from eq 2

$$\alpha_v = 1 - \frac{(\rho - \rho_a)}{\Delta\rho} \tag{2}$$

the amorphous volume fraction will be less. As the ratio between the amorphous volume fraction and the amorphous weight fraction is simply ρ/ρ_a which is independent of the choice of ρ_c, we can conclude also that the amorphous weight fraction will be less than the calculated value if $\Delta\rho$ is smaller than the ideal. Inasmuch as σ_a is equal to σ_c/α_w, if α_w is reduced, then σ_a will be increased. For example, taking $\Delta\rho$ to be 0.091, the value given by Fischer et al (10), we calculate α_w for the fiber to be 0.234 instead of 0.494 given in Table I, and σ_a for the fiber is found to be 5.39 instead of 2.56 (at 25°).

In contrast to the conclusion of Fischer et al (10) that the density of the amorphous regions in the stretched films is normal, Shimamura et al (11) who also made a study by means of low angle X-ray scattering of the density of "interlamellar" regions of drawn PE, have interpreted their data in terms of a decrease of density in the interlamellar region over that of the extrapolated melt density. Until the problem concerning the density of the amorphous regions is resolved, our estimate of σ_a for the PE fibers will be in doubt. Our data do show definitely, however, that per gram of sample the hydrogen solubility in the fiber, 1.26 p.p.m., is smaller than that of the film from the fiber, 1.51 p.p.m., although the overall densities are practically identical, 0.922 g cm^{-3} for the fiber as compared to 0.920 for the film. If the conclusion of Shimamura (11) is correct, that the density in the interlamellar regions is considerably lower than the melt density extrapolated to room temperature, one could expect a higher calculated solubility of hydrogen in the amorphous regions of the fibers than in the film, unless, of course, the degree of crystallinity was greater in the fiber.

As mentioned above the heats of solution per mole of H_2 or D_2 are considerably greater in the samples made from the branched PE than from the linear PE. This also means that the partial molal entropy of solution is also greater or in other words, the entropy of mixing or of dilution, to use the words of Jolley and Hildebrand (12), is greater in the branched PE than in the linear samples. The greater entropy of mixing in the branched PE samples suggests a greater flexibility of the chains so that more sorption sites are available.

The extended chain sample exhibits the smallest heat of solution of hydrogen. It also has the highest crystallinity with the lowest amorphous fraction. Probably the PE chains trapped in the small amorphous areas have the least flexibility of all so that the entropy of mixing is lower for these samples than for any of the others.

ACKNOWLEDGMENTS

This research was supported by income from the chair in chemistry at Baylor University endowed by a gift from The Robert A. Welch Foundation. We are indebted to Professor B. Wunderlich of Rensselaer Polytechnic Institute for the gift of the extended chain sample, to L. V. Jones of Amtech, Inc. for the fiber samples and to Dr. J. A. Reid of the Phillips Petroleum Co. for the Marlex-6002 polyethylene.

REFERENCES

1. S. Kubo and M. Dole, Macromolecules, 6, 774 (1973).

2. S. Kubo and M. Dole, Macromolecules, 7, 190 (1974).

3. W. Y. Wen, D. R. Johnson and M. Dole, Macromolecules, 7, 199 (1974).

4. V. M. Patel and M. Dole, Submitted for Publication.

5. These samples were prepared by Dr. G. N. Patel in the laboratory of Professor A. Keller of Bristol University, U.K.

6. R. Salovey, "The Radiation Chemistry of Macromolecules," M. Dole, Ed., Academic Press, New York, 1973 Vol. II, p. 308.

7. G. Herzberg, "Molecular Spectra and Molecular Structure. I. Spectra of Diatomic Molecules," D. Van Nostrand, 2nd Ed., N.Y. 1950, p. 532.

8. H. Frisch and C. E. Rogers, J. Chem. Phys. 40, 2293 (1964).

9. K. D. Ziegel and F. R. Eirich, J. Poly. Sci. Part A-2, 8, 2015 (1970).

10. E. W. Fischer, H. Goddar and G. F. Schmidt, Macromol. Chem. 118, 144 (1968).

11. K. Shimamura, S. Murakami and K. Kobayashi, Bull. Inst. Chem. Res., Kyoto University, Japan, 52, 359 (1974).

12. J. E. Jolley and J. H. Hildebrand, J. Am. Chem. Soc. 80, 1050 (1958).

PHASE SEPARATION BEHAVIOR OF SODIUM POLYACRYLATE AND SODIUM POLYSTYRENESULFONATE IN THE PRESENCE OF BARIUM CHLORIDE

Paul L. Dubin
Dynapol

I. INTRODUCTION

Virtually all polyelectrolytes are susceptible to precipitation in the presence of multi-valent counterions. This phenomenon clearly arises from the binding of counterions to the polyions. The interaction of divalent counterions with polyanions has been rather thoroughly investigated by techniques such as potentiometry (1-5), dialysis equilibrium (6), dilatometry (6-10), difference spectroscopy (1) and calorimetry (9,10), and the phenomenon of site-binding -- the formation of ion-pairs with the release of water of solvation -- has been conclusively demonstrated (7). These studies always stop short of phase separation which complicates both the measurements and interpretation of results. Thus, relatively few authors have dealt with the precipitation of polyelectrolytes resulting from divalent ion-binding (11,12).

The scarcity of literature does not however imply a lack of relevance of this issue. It is not difficult to see that many applications of synthetic polyelectrolytes could be limited by such behavior. A closely related problem, for example, is the selection of detergents for hard water systems since typical ionic surfactants are salted-out in the presence of calcium ion (13).

Several mechanisms may be envisaged for the precipitation of polyanions by divalent cations, as illustrated in Figure 1. In the first model, the aqueous affinity of ionic sites, and hence the solubility of the polymer, is lost upon electrical neutralization. Phase separation then occurs when a given fraction of ionic sites is occupied by bound metal ion. This simple model predicts that the concentration of divalent ion at phase separation increases in a linear manner with polymer concentration (11). If some or all of the counterions form ionic bridges between groups far apart on the polymer chain, as in Model II, solute concentrations are still related in a similar manner, although one might expect the degree of binding at phase separation to be lower. Similar ionic cross-linking leads to intermolecular association in Model III. In this case a "solubility-product" type relationship holds, and the critical concentrations of polymer and counterion vary inversely, in contrast to the behavior predicted for the first two models.

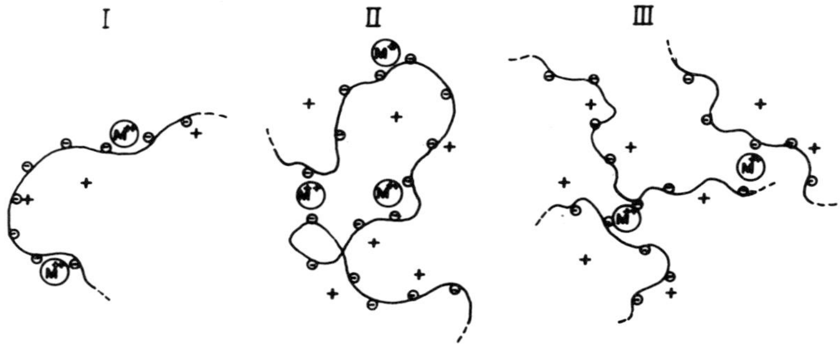

$$\bar{C}_S = \beta C_p + \gamma \qquad\qquad C_S^m C_p^n = K$$

Fig. 1. Models for polyanion-cation complex at incipient phase separation. See text for explanation.

The "site neutralization" mechanism (Model I) was found to readily accommodate the data for calcium, barium and magnesium polymethacrylate (11), magnesium polyphosphate (14), and barium polystyrenesulfonate (12). Thus, these studies provide no evidence for intermolecular bridging. However, such interactions are thought to be significant at least in the case of anionic polysaccharides, in particular gels of alginates and pectins that contain calcium or magnesium (15). The current work was motivated by the expectation that phase separation studies with narrow-distribution fractions of widely varying molecular weights might help discriminate among the various proposed models. It is primarily in this regard that the present study differs from previous ones with polymethacrylate (11) or polystyrenesulfonate (12).

II. EXPERIMENTAL

Narrow MWD samples of sodium polystyrenesulfonate (NaPSS) were prepared by the method of Carroll and Eisenberg (12). Exhaustive dialyses against 0.1 N HNO_3, 1 M NaCl and pure water, in succession, were required to remove colloidal and ionic silver arising from the catalyst employed in sulfonation of the narrow MWD polystyrene samples (Pressure Chemical Company). Narrow MWD fractions of sodium polyacrylate (NaPAA) were prepared by solvent-non-solvent fractionation (16) of commercial polyacrylic acid samples, followed by neutralization with NaOH and freeze-drying. The characteristics of the NaPSS and NaPAA fraction are given in Table 1.

TABLE 1. *Characteristics of Polymer Fractions of this Study.*

CODE	MW	$MW/\bar{M}_n$
NaPSS S10	1.4×10^{6} [a]	
NaPSS S 9	2.2×10^{5}	
NaPSS S 8	7.3×10^{4}	< 1.1
NaPSS S 7	4.0×10^{4}	
NaPSS S 6	2.0×10^{4}	
NaPSS S 5	4.2×10^{3}	
NaPAA A 6	4.8×10^{5} [b]	< 1.3
NaPAA A 5	4.6×10^{4}	
NaPAA A 4	6.8×10^{3}	
NaPAA A 3	2×10^{3}	

a. From molecular weight of parent polystyrene.
b. From intrinsic viscosity (17).

Turbidimetric titrations were carried out at temperatures ranging from 15-65°C in a water bath (±0.5°C), using 0.5 ml portions of polymer solutions. A 0.2 ml micrometer syringe buret was used to add 1.00 M $BaCl_2$. An intense collimated beam of light illuminated the sample. The value of $\bar{C}_s$, the lowest molar concentration of divalent ion at which turbidity could be detected, was reproducible to within 5%. In the case of NaPSS, conditions for phase separation could also be established by slowly cooling a heated solution of BaPSS and the values so obtained were in agreement with the results of turbidimetric titrations.

Reduced viscosities were determined with a Sofica "Viscomat" automatic viscometer thermostated at 25.0°C. GPC measurements were carried out with an instrument of our design, comprised of a high pressure pump and injector (Milton Roy, Valco), columns packed with CPG® porous glasses, and a variable wavelength UV detector (Schoeffel) set at 220 nm.

All measurements were made with 0.1 M NaCl as supporting electrolyte. In some cases (i.e., low MW NaPSS), the ionic strength contribution of $BaCl_2$ approached or even exceeded this value. It was established, however, that molar concentrations of NaCl failed to lead to precipitation and the ionic strength component of the observed phase separations is thus of secondary importance. On the other hand, high concentrations of NaCl can reduce the effectiveness of Ba^{++} as a precipitant, presumably *via* competitive binding.

III. RESULTS

A. Turbidimetric Titrations

NaPSS

Turbidimetric titrations were carried out at polymer concentrations of C_p = 0.31, 0.57, 1.0, 2.0 and 4.0 mg ml^{-1}, at temperatures ranging from 17°-65°. Representative plots of T_p *vs.* $\bar{C}_s$ (the temperature and the concentration of Ba at phase separation, respectively) are shown in Figure 2 for

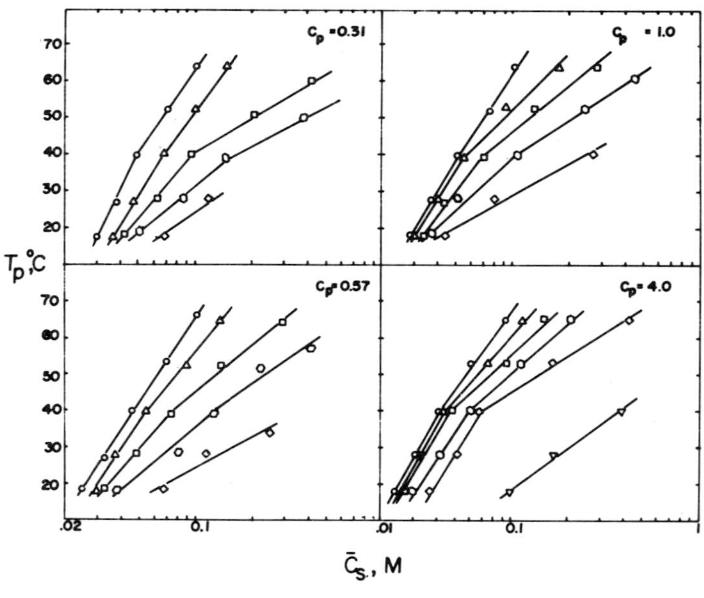

Fig. 2. Precipitation temperature vs $BaCl_2$ concentration: (○) S-10, (△) S-9, (□) S-8, (⬡) S-7, (◇) S-6, (▽) S-5.

C_p = 0.31, 0.57, 1.0 and 4.0. For the intermediate value of C_p = 2.0 qualitatively similar plots were obtained. Linear dependences of T_p on log $\bar{C}_s$ were observed at $T_p \leq 40°$; the marked change in slope at $T_p \cong 40°$ was characteristic of all such plots regardless of C_p or MW.

The results of turbidimetric titrations may also be presented as plots of $\bar{C}_s$ *vs.* C_p at constant T_p. The data points of Figures 3 and 4 correspond either to actual measured values or to interpolation from the plots in Figure 2.

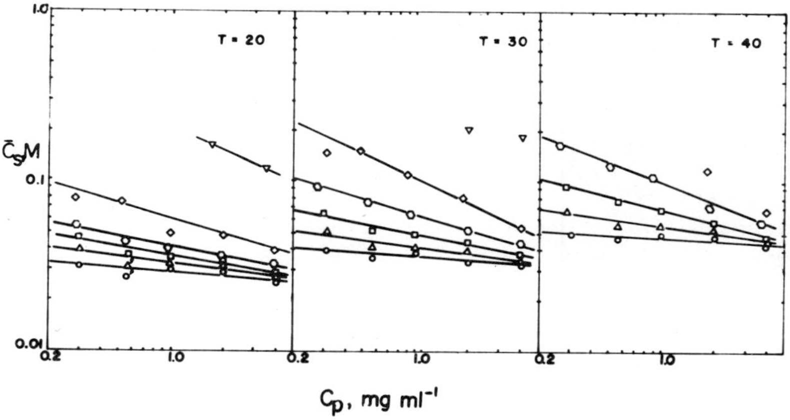

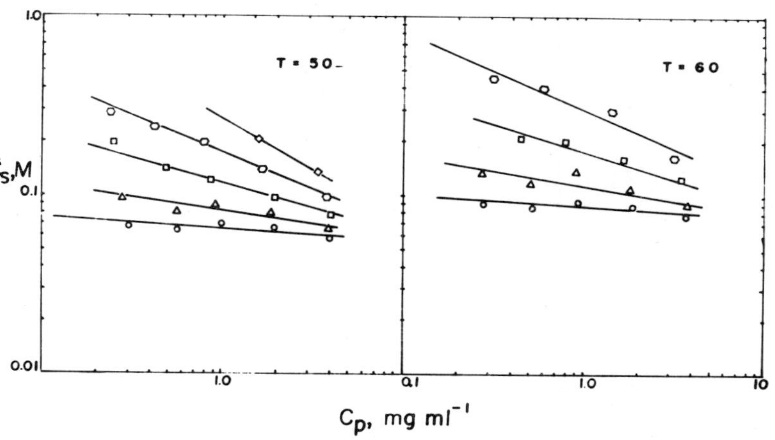

Figs. 3 and 4. Dependence of $BaCl_2$ concentration at phase separation on polymer concentration. Symbols as for Fig. 2.

NaPAA

In marked contrast to BaPSS, the precipitation of BaPAA was rather insensitive to temperature, especially for fractions A6 and A5. Somewhat surprisingly, the lower MW fractions displayed a modest inverse temperature dependence, i.e., $\bar{C}_s$ decreasing as the temperatures increased.

The dependence of $\bar{C}_s$ on polymer concentration for NaPAA is shown in Figure 5. Also included are data of Michaeli (11) obtained under similar conditions for a sample of sodium polymethacrylate of unspecified MW. As in the case of NaPSS, a linear dependence of log $\bar{C}_s$ on log C_p is observed. On the other hand, the concentration of Ba^{++} at phase separation is greater by a factor of four or more for NaPSS, at comparable conditions of polymer normality, degree of polymerization and temperature. The most striking qualitative contrast between the behavior of two types of polymers is evident from the slopes of Figure 5 *vs.* those of Figures 3 and 4. While the amount of Ba^{++} required for precipitation of NaPAA increases with increasing polymer concentration, inverse dependence of $\bar{C}_s$ on C_p is seen for NaPSS.

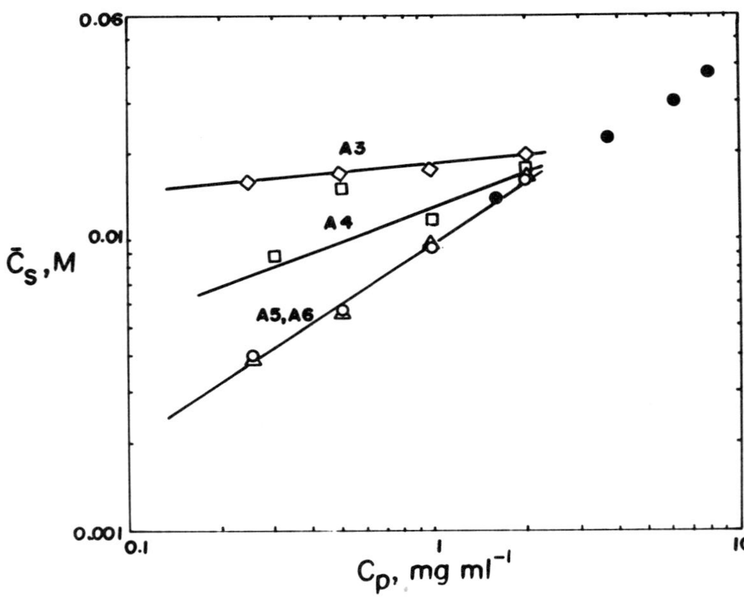

Fig. 5. Dependence of $BaCl_2$ concentration at phase separation on polymer concentration for NaPAA at 25°C. (○) A6, (△) A5, (□) A4, (◇) A3. ●: Data for NaPMA from Ref. 11.

B. Viscometry and GPC - NaPSS

The influence of Ba^{++} on viscosity should reflect the way in which binding alters molecular dimensions. The dependence of η_{sp}/C_p for S10 is given in Figure 6. The viscosities of the other polymers were too small to allow application of this method in the range of polymer concentrations of interest.

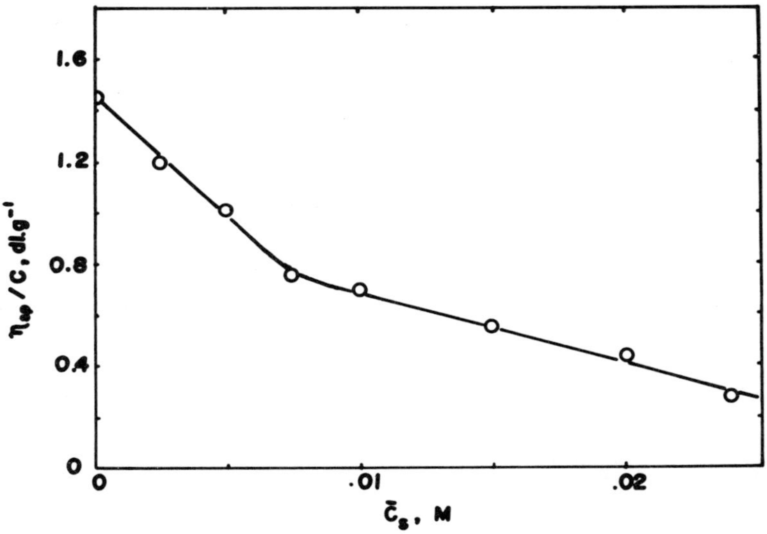

Fig. 6. Effect of $BaCl_2$ concentration on the viscosity of S-10 at 25°C. C_p = 2.0 mg ml^{-1}.

Molecular dimension at very low polymer concentrations may, however, be measured by GPC. Samples S5-10 were injected at polymer concentrations of 0.3 mg ml^{-1} and at Ba^{++} concentrations somewhat less than those required for phase separation. Since the values of C_p and C_s may be presumed to diminish during elution through the column, the conditions of this experiment are not well defined. In Figure 7 the apparent molecular weights of the BaPSS samples, established by reference to a NaPSS calibration curve, are given as a function of MW. Despite the uncertainties noted above, the results clearly indicate reduction in molecular size upon addition of $BaCl_2$.

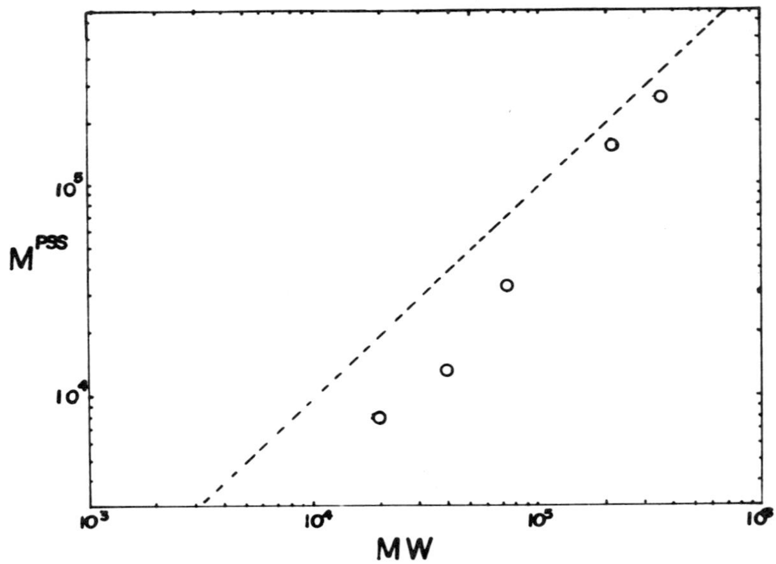

Fig. 7. GPC molecular weights (relative to NaPSS) vs true molecular weights. Broken line is M^{PSS} = MW.

IV. DISCUSSION

The results obtained from the turbidimetric titrations of NaPAA are generally consistent with findings for other polyanions. As shown in Figure 5, the dependence of $\bar{C}_s$ on C_p is in harmony with that found for NaPMA (11). Indeed, the data for NaPAA -- although plotted bi-logarithmically in Figure 5 for comparison with Figures 3 and 4 -- conform as well to a linear non-logarithmic relationship, consistent with the expression proposed by Michaeli (11):

$$\bar{C}_s = \beta N_p + \gamma \tag{1}$$

In this interpretation βN_p and γ are the molar concentrations of free and bound divalent cation, precipitation occurring when the fraction of sites occupied is β. A similar expression was proposed by Strauss on the basis of data for magnesium polyphosphate (14). The requirement for higher $BaCl_2$ concentration to precipitate lower MW polymer, evident in Figure 5, is in accord with expectations. But if the data of Figure 5 are plotted according to eq. (1), one obtains

β= 0.7, 0.7, 0.5 and 0.2 for fractions A6, A5, A4 and A3, respectively. This result is surprising since it would be expected that the fraction of sites that must be occupied for precipitation should <u>increase</u> as the MW decreases.

Turning to the phase separation behavior of NaPSS, one finds a dependence of $\bar{C}_s$ on C_p evidently inconsistent with eq. (1).[1] (It may be noted here that semi-logarithmic or non-logarithmic plots of the data of Figs. 3 and 4 do not give straight lines.) The observed inverse relation between the critical concentrations of Ba^{++} and NaPSS is clearly discordant with Models I and II previously described. Model III, on the simplest level, may be expressed as the following equilibrium:

$$n\ M^{++} + m\ P \rightleftarrows M^{++}_n\ P_m \downarrow \qquad (2)$$

with

$$K_{sp} = \bar{C}_s^{\ n}\ C_p^{\ m} \qquad (3)$$

or

$$\log \bar{C}_s = -\frac{m}{n} \log C_p + \log K_{sp}^{\ 1/n} \ . \qquad (4)$$

The data for NaPSS are seen in Figures 3 and 4 to conform to eq. (4). The value n/m is the stoichiometric ratio of divalent ions and polymer molecules that interact to form an insoluble aggregate, and is given by the negative reciprocal of the slope in a plot of log $\bar{C}_s$ *vs* log C_p. The number of divalent ions bound per polymer molecule at phase separation might be expected to increase with polymer chain length, so that $n/m = -(d \log \bar{C}_s/d \log C_p)^{-1}$ would be proportional to the degree of polymerization DP. In Figure 8, $-(d \log \bar{C}_s/d \log C_p)$, the negative of the slopes of Figures 3 and 4 is found in fact to vary linearly with 1/DP for $T_p \leq 40°$.[2]

---

[1] In contrast to these results, Carroll and Eisenberg (12) found a direct linear dependence of $\bar{C}_s$ on C_p for a sample of NaPSS of MW = 9×10^5. While there are some differences in experimental conditions -- higher polymer concentrations and an absence of supporting electrolyte in Ref. 12 -- there is at present no satisfying explanation for the discrepancy between the present results and those of Ref. 12.

[2] A general thermodynamic treatment by Michaeli (18) for the incipient precipitation of polymer by stoichiometric or non-stoichiometric interaction with species i leads to: $(\partial \ln C_p/\partial \ln a_i)_{\mu_j} = (DP)\Gamma_i$, where $\Gamma_i = (\partial C_p/\partial \mu_i)_{\mu_j}$. When C_i/N_p is large, as in the present case, Γ_i is relatively insensitive to C_p(18). Thus, a very general treatment seems to predict the behavior shown in Fig. 3 & 4 and Fig. 8. It is not clear however whether these relations can rationalize the contrasting behavior of NaPAA and NaPSS.

The change in this dependence at $T_p > 40°$, taken with the discontinuities of Figure 2, suggest a change in the mechanism of phase separation at *ca* T = 40°.

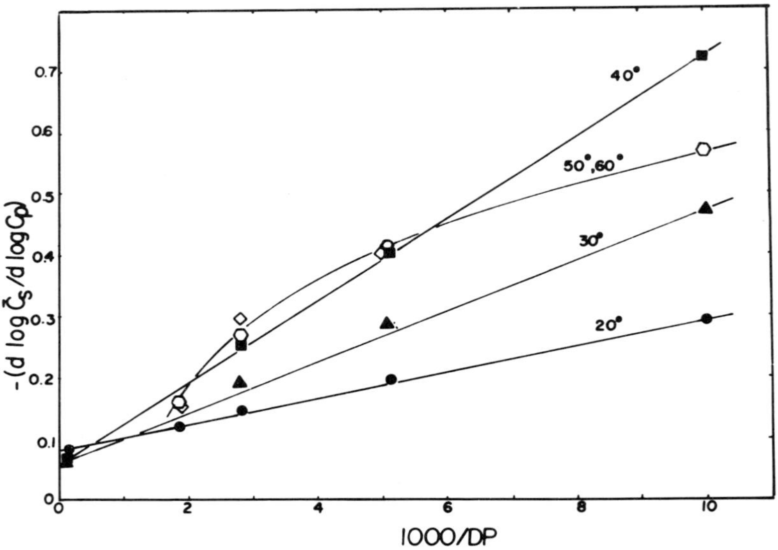

Fig. 8. Negative slopes of Figures 3 and 4 vs reciprocal degree of polymerization.

The foregoing analysis might seem to imply that n and m are constants depending only on temperature and polymer molecular weight. In order to exclude the multiple equilibria that correspond to a range of values of n and m, a highly specific and cooperative process would be required. It would seem more reasonable to assume only that the average ratio of n/m is fairly insensitive to $\bar{C}_s/C_p$ for any given polymer fraction at a constant temperature.

The process put forth in eq. (2) corresponds at least in part to the intermolecular association model (III) in Fig. 1. Such aggregation might be expected to lead to an increase in polymer dimensions. The viscosity and GPC results of Figs. 6 and 7, respectively, indicate molecular contraction prior to phase separation. However, the precipitation mechanism described by eq. (2) need not exclude simple site-binding, *i.e.*, only a fraction of bound divalent cation may be involved in intermolecular bridges. The decrease in electrostatic expansion due to site neutralization may dominate any molecular expansion from aggregation.

At fixed solute concentrations, T_p increases strongly with MW, as may be seen by comparison of the appropriate data points of Figure 2. This dependence may arise from a number of effects, including changes in the precipitation mechanism. Figure 9 presents the MW-dependence of T_p at $\bar{C}_s = 0.1$, plotted according to the theory for polymer phase-separation in binary systems (19). The data are in good agreement with the predicted dependence of T_p^{-1} on $MW^{-\frac{1}{2}}$, although it is not clear whether the congruence of the data with a theory developed for a much simpler case is fortuitous.

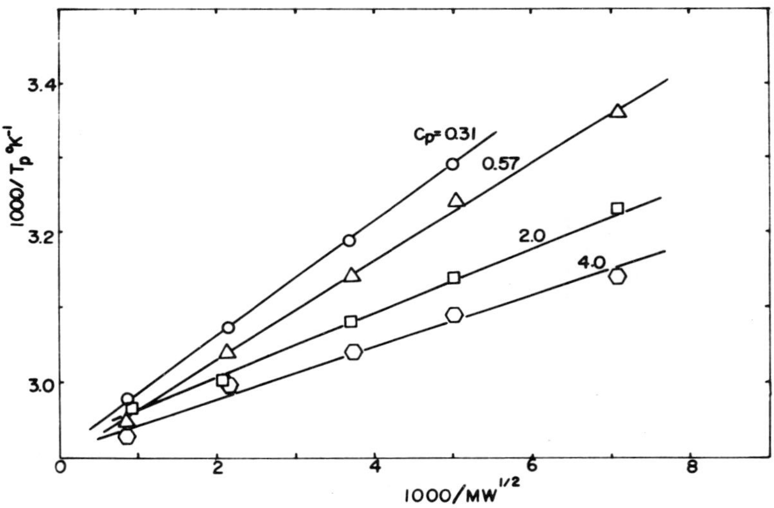

Fig. 9. Molecular weight dependence of T_p for various polymer concentrations at $\bar{C}_s = 0.1$, plotted according to theory of Flory (19).

The present results reveal dramatic differences in the phase separation behavior of polyacrylate and polystyrenesulfonate. These contrasting features are summarized in Table 2. The two polymers also display very different dilatometric behavior with divalent cations: for polyacrylate, as for polyphosphate, typical values of ΔV range from 15-25 ml/mole divalent ion (7,8), while the values for polystyrenesulfonate are on the order of only 2-3 ml (7). The interaction of divalent cations with NaPAA probably involves in large part the formation of strong, desolvated ionic associations between the metal ion and adjacent pairs of carboxylates (7,8). Both the bulkiness and the spacing of the sulfonate groups in NaPSS make such association difficult (7). The weakness of the M^{++} - PSS^{-} site interaction presumably diminishes the role of site neutralization in precipitation and enhances the importance of longer range interactions. In this regard, the large volume of the barium ion renders it most effective in the formation of intermolecular or long-range intramolecular ionic bridges.

TABLE 2. *Comparison of Phase Separation of Polyacrylate and Polystyrenesulfonate in the Presence of Divalent Cations.*

Polyacrylate	Polystyrenesulfonate
1. Concentration of M^{++} required for phase separation increases with polymer concentration.	1. Concentration of M^{++} at phase separation decreases with increasing polymer concentrations.
2. Mg^{++}, Ca^{++} and Ba^{++} behave identically in precipitating NaPMA (11).	2. Only Ba^{++} causes precipitation.
3. Typical ratio of normalities of M^{++} and polymer at phase separation *ca* 1.	3. Typical ratio of critical Ba^{++} normality to polymer normality at 25° *ca* 2-20.
4. Weak or "inverse" temperature dependence.	4. Strong "normal" temperature dependence.

Further experiments may establish whether other metal ion-polyion systems exhibit the apparently atypical behavior of barium-polystyrenesulfonate found in the present study.

ACKNOWLEDGMENT: The author wishes to thank D. Matzinger and A.R. Cooper for the samples of NaPSS. Helpful discussions with M. Kronstadt, W.J. Leonard, Jr. and U.P. Strauss are gratefully acknowledged.

REFERENCES

1. F.T. Wall and S.J. Gill, *J. Phys. Chem.*, *58*, 1128 (1954).
2. M. Mandel and J.C. Leyte, *J. Polymer. Sci.*, *2*, 2883 (1964).
3. B.J. Felber, E.M. Hodnett and N. Purdie, *J. Phys. Chem.*, *72*, 2496 (1968).
4. M. Rinaudo and M. Milas, *J. Polymer Sci.*, *12*, 2073 (1974).
5. C. Travers and J.A. Marinsky, *J. Polymer Sci.*, Symp. No. 47, 285 (1974).
6. J. Hen and U.P. Strauss, *J. Phys. Chem.*, *78*, 1013 (1974).
7. U.P. Strauss and Y.P. Leung, *J. Amer. Chem. Soc.*, *87*, 1476 (1965).
8. A.J. Begala and U.P. Strauss, *J. Phys. Chem.*, *76*, 254 (1972).
9. V. Crescenzi, F. Delben, S. Paoletti and J. Skerjanc, *J. Phys. Chem.*, *78*, 607 (1974).
10. F. Delben and S. Paoletti, *J. Phys. Chem.*, *78*, 1486 (1974).
11. I. Michaeli, *J. Polymer Sci.*, *48*, 291 (1960).
12. W.R. Carroll and H. Eisenberg, *J. Polymer Sci.*, *4*, 599 (1966).
13. K. Shinoda and T. Hirai, paper presented at the 7th Northeast Regional Meeting, ACS, 1976.
14. U.P. Strauss and A. Siegel, *J. Phys. Chem.*, *67*, 2683 (1963).
15. R. Speiser, C.H. Hills and C.R. Eddy, *J. Phys. Chem.*, *49*, 328 (1945).
16. S. Newman, W.R. Krigbaum, C. Laugier and P.J. Flory, *J. Polymer Sci.*, (1954).
17. A. Takahashi and M. Nagasawa, *J. Amer. Chem. Soc.*, *86*, 543 (1964).
18. I. Michaeli, *J. Phys. Chem.*, *71*, 3384 (1967).
19. See, for example, P.J. Flory, *Principles of Polymer Chemistry*, Cornell University Press, Ithaca, N.Y., 1953, Ch. XIII.

CONFORMATION OF SULFONATE-CONTAINING POLYMERS

L. W. Fisher and J. S. Tan
Research Laboratories, Eastman Kodak Company

I. INTRODUCTION

Previous theoretical treatments of chain flexibility and chain expansion have been based on the assumption that chain conformation and the energies of rotational isomeric states at a theta condition are insensitive to the nature of solvent media or to the degree of ionization of the charged groups.[1,2] Although this assumption has been shown to be approximately valid for many nonionic homopolymers in single nonpolar solvents,[1,3] numerous exceptions have also been observed. Different unperturbed dimensions or chain flexibilities have been reported by several workers for nonionic homopolymers in single solvents.[3-7] Additionally, these dimensions have been found to vary with solvent composition for nonionic polymers[8-13] as well as for copolymers[14,15] in binary mixed solvents. These variations in unperturbed dimensions have been ascribed to preferential adsorption and have been interpreted as a function of thermodynamic interaction parameters of the solvent pair.

Studies of polyelectrolytes in various solvents are limited in number. Frisman and coworkers[16] have reported considerable variations of the unperturbed dimensions of poly(methacrylic acid) (PMAA) as a function of methanol-water composition. Flory and Osterheld[17] have reported identical unperturbed dimensions for poly(acrylic acid) (PAA) in dioxane and for its sodium salt in aqueous NaCl solutions. The extensive work of Takahashi and Nagasawa[18] has shown that fully ionized poly(acrylic acid) has much larger unperturbed dimensions in aqueous NaBr solutions than those of PAA in dioxane. Differences in the unperturbed dimensions have also been observed for PMAA in methanol and in 0.002 $\underline{N}$ HCl solutions.[19-21] Our work on a copolymer of acrylic acid and ethyl acrylate[22] also has shown great differences in the polymer's unperturbed dimensions in organic and in aqueous NaCl solutions. In light of the review paper by Morawetz,[23] it is theoretically correct to expect variation of the unperturbed dimensions of polyelectrolytes with the degree of ionization and with the ionic strength of the medium.

Very little work has been published on the effect of counterions on the chain flexibility of polyelectrolytes. Based on the results for potassium poly(vinylsulfonate) (K-PVS),[24] potassium poly(styrenesulfonate) (K-PSS),[25] and sodium poly(styrenesulfonate) (Na-PSS),[26] it has been assumed that unperturbed dimensions are slightly sensitive to the counterions and insensitive to the coions or the theta temperature.

Fractions of poly(2-acrylamido-2-methylpropanesulfonate) (PAMS) have been characterized by us through light-scattering and intrinsic-viscosity ([η]) measurements.[27,28] The sodium salt form of this polymer (Na-PAMS) is soluble only in aqueous media. The acid form (H-PAMS) exhibits solubility in aqueous media and is also soluble in a limited number of highly polar organic solvents such as methanol and dimethylformamide. The tetrabutylammonium (TBA) salt form is soluble in a wide range of organic solvents, e.g., isobutyl alcohol, dichloroethane, and acetone. In this study, we compare the unperturbed dimensions of TBA-PAMS in organic and in aqueous NaCl solutions. Further comparisons are drawn for the following PAMS salts in aqueous salt solutions: TBA-PAMS in TBA-Br and in NaCl; Na-PAMS in NaCl, TBA-Br, and CsCl; H-PAMS in NaCl; and Cs-PAMS in NaCl and CsCl. Our present data demonstrate further the effect of (1) the nature of the solvent media and (2) the nature of counterions on the unperturbed dimensions for this sulfonate-containing polymer.

Several water-soluble sulfonate-containing polymers were also studied by intrinsic-viscosity and fluorescence probe techniques in order to elucidate the relationship between polymer conformation and solubility behavior. Polymers studied were PAMS; PVS; PSS; poly(3-methacryloyloxypropane-1-sulfonate) (PMOS); and three copolymers of N-t-butylacrylamide and 2-acrylamido-2-methylpropane sulfonate (NB-AMS) (mole ratios 1.1:1, 2.3:1, and 4.4:1). The fluorescence probe, Na-6-p-toluidinonapthalene-2-sulfonate (TNS), first used to study protein structures,[29,30] was utilized to monitor changes in conformation of these polymers in aqueous solutions as a function of ionic strength. Previous binding studies[31] with this dye have revealed a conformational transition, for the above-mentioned copolymer of acrylic acid and ethyl acrylate, from a highly extended coil at low ionic strength to a compact coil at high ionic strength. The compact form of the polymer coil is thought to arise from shielding of ionic groups on the polyelectrolyte by the added electrolyte in conjunction with clustering of hydrophobic portions of the polymer molecules.[31] The TNS probe, as will be discussed below, has been found to be sensitive to the extent of this hydrophobic clustering.

II. EXPERIMENTAL

A. Materials

The preparations of Na-PMOS[32] and its fractions ($\bar{M}_w$ = 1.05×10^5 to 2.75×10^6) and that of H-PAMS[27] and its fractions ($\bar{M}_w = 1.27 \times 10^5$ to 7.84×10^5) have been described elsewhere. The various salts of PAMS were prepared by neutralization of the acid form of the polymer with the corresponding hydroxide. The preparations of NB-AMS and PSS have also been described previously.[27] Fractionation of the NB-AMS copolymers was carried out by stepwise precipitation. A solvent-non-solvent system of water-dioxane was used to fractionate NB-AMS (1.1:1) while ethanol was used as the solvent and heptane as the non-solvent to fractionate the other two NB-AMS polymers. Weight-average molecular weights were determined for the NB-AMS fractions used in this study by light-scattering using a Sofica photometer at 436 nm following a dialysis period of 2-4 days.

The H-PVS polymer was prepared by R. C. Sutton of these Laboratories according to the method suggested by Eisenberg and Mohan.[33] In the stepwise fractionation of this material a solvent-nonsolvent system of water-methanol was used. The Na-PSS equivalent weight-average molecular weight ($\bar{P}_w$) of the fraction used here was determined by aqueous exclusion chromatography.

The TNS solution was prepared as previously described.[31] Values of $\bar{M}_w$ ($\pm$ 5%) for the fractions used in the fluorescence experiments are as follows: Na-PAMS, 235,000; Na-PMOS, 660,000; Na-PSS, 400,000; Na-PVS ($\bar{P}_w$), 93,000; Na-NB-AMS (1.1:1), 140,000; Na-NB-AMS (2.3:1), 410,000; Na-NB-AMS (4.4:1), 160,000.

B. Methods

1. *Intrinsic Viscosity Measurements*

Intrinsic-viscosity measurements were carried out in Ubbelöhde semi-micro viscometers at 25.0° $\pm$ 0.05°C. Polymer solutions and solvents were passed through either 5 μm Teflon Millipore filters or fine-porosity sintered-glass filters prior to use. No kinetic energy corrections were made on the viscosity data since all flow times exceeded 100 sec. No shear-rate dependence of viscosity was noted in checks of a number of high molecular weight fractions at low ionic strengths in aqueous media.

2. *Fluorescence Intensity Measurements*

Solutions at various NaCl concentrations were prepared at 0.5 percent polyelectrolyte containing 1.6 x 10^{-5} <u>M</u> TNS, and were held at 25°C for 30 minutes before measurement of fluorescence. The experimental setup for fluorescence intensity determinations was designed by Dr. G. L. Beyer, Mr. L. F. Costa, and Mr. T. H. Prevost of these Laboratories. Cylindrical cells containing the solutions were centered in a hexagonal cell and thermostated at 25°C. A 100 watt (H-4) mercury lamp was used as the light source with the incident light passing through a Kodak #18A glass filter (maximum transmittance at 365 nm). Emitted light was monitored by a photomultiplier (1P28) positioned at 45° relative to the incident beam. Light of the exciting wavelength was eliminated through use of a Corning #4305 glass filter in combination with Kodak Wratten filters #2A and #47 (maximum transmittance at 480 nm, zero transmittance at 365 nm). The emission maximum of the TNS probe,[8] occurs at ~440 nm. The output of the photomultiplier was monitored on a strip chart recorder until a constant value was reached. In some cases, samples were mixed to obtain solutions of intermediate ionic strengths. A few minutes was found to be adequate time for re-equilibration of these solutions (as indicated by constant fluorescence intensity).

A 10^{-4} M quinine sulfate solution (in 0.1 N H_2SO_4), used as a reference to compensate for day-to-day fluctuations in output of the light source, was prepared from quinine sulfate purified by recrystallization from methanol prior to use. The fluorescence intensities of NaCl solutions containing TNS but no polymer were measured to correct for variations in emission with ionic strength.

III. RESULTS AND DISCUSSION

A. Unperturbed Dimensions of PAMS Salts

Table I lists the intrinsic-viscosity data for TBA-PAMS in i-BuOH and $C_2H_4Cl_2$. In both cases, the measurements were carried out in the presence of 0.001 M TBA-Br to suppress the repulsive effect of polyelectrolytes upon dilution. Values of $[\eta]$ are also given for the various PAMS salts in aqueous salt solutions. Estimated errors in these values are ± 2%.

Among various $[\eta]$-M relationships available in the literature, the Stockmayer-Fixman extrapolation technique[34] represents the most convenient and commonly used method for determining the unperturbed dimensions. This equation is expressed as follows:

$$[\eta]/M^{1/2} = K_o + 0.51\,\Phi_o\, BM^{1/2} \qquad (1)$$

where K_o is the intercept and is equal to the Mark-Houwink constant K at a theta condition, B is related to the second virial coefficient, and Φ_o is the limiting Flory viscosity constant. The value K_o is related to the characteristic ratio, C_∞, and the unperturbed end-to-end distance, $\langle r_o^2\rangle$, by[1]

Table I. Intrinsic-Viscosity Data for

Fraction	$\overline{M}_w \times 10^{-6}$ (H-PAMS)	H-PAMS in 5N NaCl	Na–PAMS in NaCl 0.01N	0.05N	0.1N	0.5N	1.0N	5.0N
F1	0.78_4	1.0_7	7.1_7	3.5_3	2.8_0	1.6_8	1.3_6	1.0_2
F2	0.49_5	0.78_5	4.8_1	2.4_9	1.9_0	1.1_5	0.98_0	0.74_7
F3	0.37_2	0.59_5	3.4_6	1.7_8	1.4_1	0.88_1	0.69_5	0.53_7
F4	0.20_2	0.39_2	1.8_2	1.0_2	0.84_5	0.52_0	0.46_7	0.40_4
F5	0.12_7	0.25_0	1.1_5	0.67_2	0.53_2	0.36_4	0.32_6	0.25_5

$$C_\infty = \left(\frac{K_o}{\Phi_o}\right)^{2/3}\left(\frac{M_o}{2\ell^2}\right) = \frac{\langle r_o^2\rangle}{n\ell^2} \qquad (2)$$

where M_o is the monomer molecular weight, ℓ is the bond length, and n is the total number of backbone bonds.

Plots of $[\eta]/M^{1/2}$ vs. $M^{1/2}$, according to equation (1), are shown in Figure 1 for Na-PAMS in NaCl solutions, in Figure 2 for TBA-PAMS in the various solvent media, and in Figure 3 for the remaining systems investigated. Values of K_o obtained from these plots and the calculated values of C_∞ ($\pm$ 10%) are listed in Table II. For Na-PAMS in aqueous NaCl solutions and for TBA-PAMS in NaCl, the values of K_o prove to be independent of ionic strength. This independence is assumed to hold for the other systems investigated. A common intercept is also realized for TBA-PAMS in both organic solvents.

Within experimental error, three distinct C_∞ values were obtained, reflecting three degrees of flexibility of the PAMS coil in its unperturbed state. Apparently, the TBA ions are only weakly (diffusely) bound to sulfonate groups on the polymer. These ions contribute to the extension of the polymer chain due to their bulkiness and are relatively ineffective in screening electrostatic repulsions among sulfonate groups. The higher value of C_∞ for TBA-PAMS in TBA-Br relative to TBA-PAMS in organic media may be attributed to differences in electrostatic interactions, the polymer being fully ionized in the former case and largely unionized in the latter.

A common value for C_∞ was obtained for the remaining systems, all of which contain ions which are more strongly

PAMS Salts in Various Solvent Media

Na-PAMS in 1N CsCl	Na-PAMS in 0.5N TBA-Br	Cs-PAMS in 1N CsCl	Cs-PAMS in 1N NaCl	TBA-PAMS in NaCl		TBA-PAMS in 0.5N TBA-Br	TBA-PAMS in i-BuOH	TBA-PAMS in $C_2H_4Cl_2$
				1.0N	5.0N			
1.6_2	1.3_6	1.0_3	0.90_7	0.79_0	0.55_5	1.4_5	2.87	1.0_6
1.1_1	0.94_8	0.70_8	0.70_5	0.57_2	0.38_5	1.0_0	1.9_0	0.73_5
0.85_3	0.72_5	0.56_2	0.51_0	0.42_0	0.28_5	0.77_4	1.4_5	0.51_7
0.50_0	0.39_3	0.30_3	0.28_0	0.25_7	0.16_5	0.48_2	0.81_2	0.33_3
-	0.29_5	-	-	0.16_5	0.09_5	0.32_1	0.49_0	0.23_5

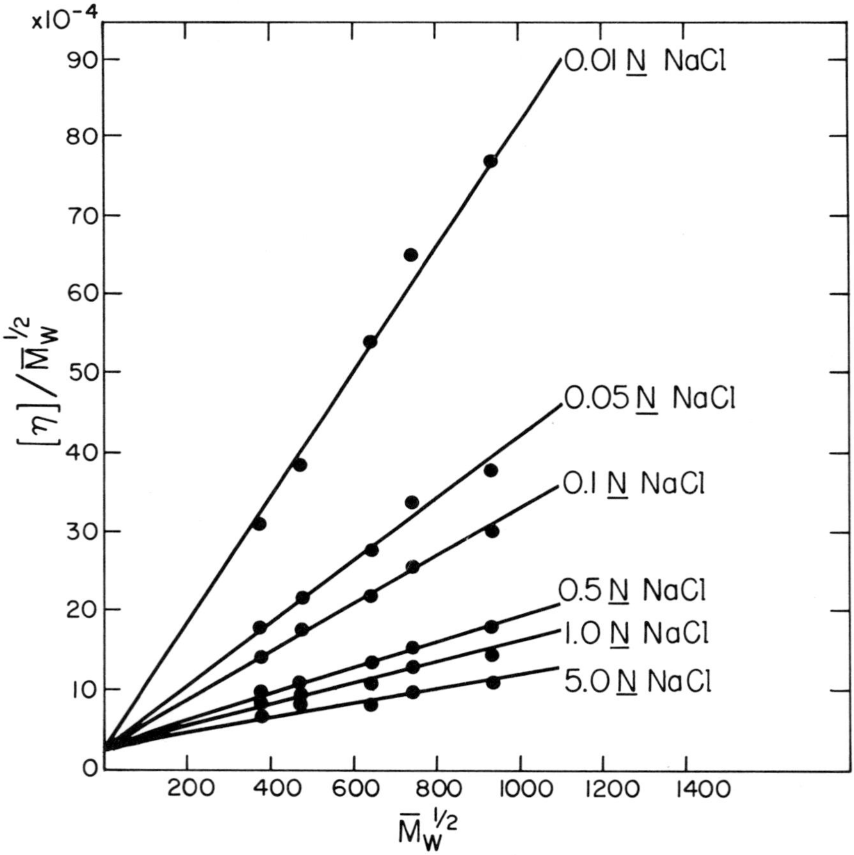

Fig. 1. The Stockmayer-Fixman plots for Na-PAMS in aqueous NaCl solutions.

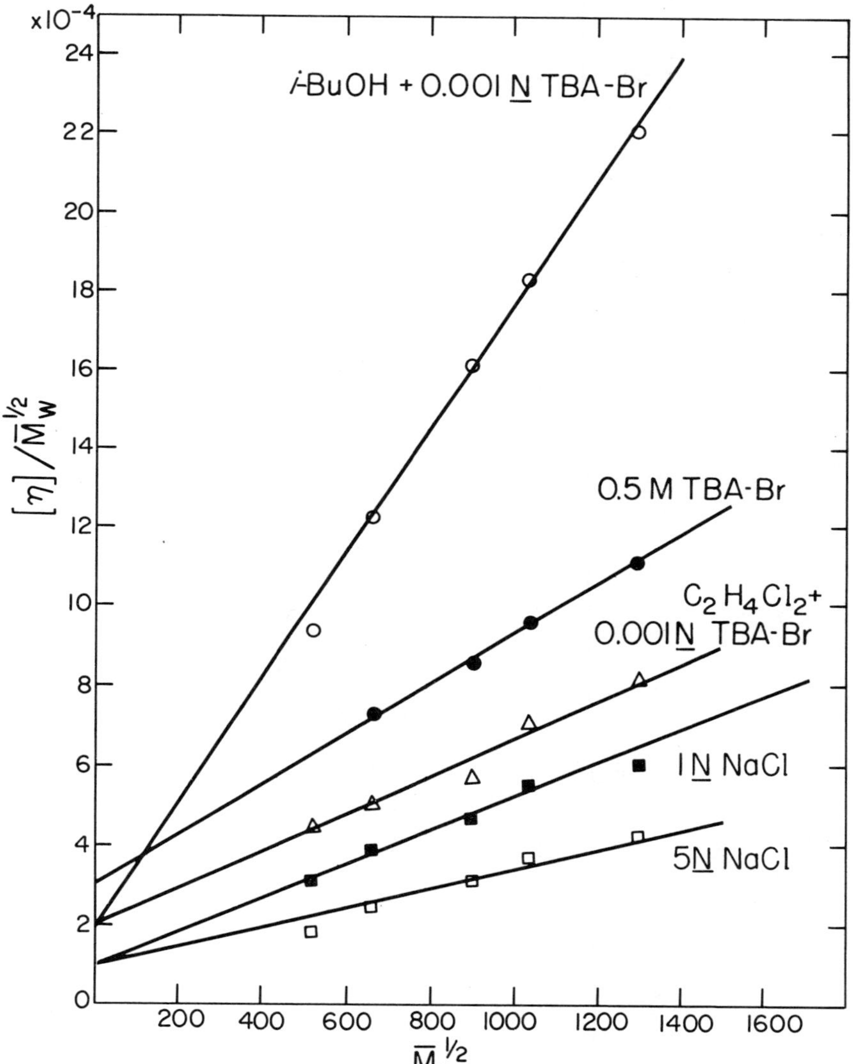

Fig. 2. The Stockmayer-Fixman plots for TBA-PAMS in organic and in aqueous media.

bound to the polymer. Within experimental error H^+, Na^+, and Cs^+ have an equal effect on determining the unperturbed dimensions of the PAMS coil. In the case of TBA-PAMS in NaCl, Na^+ ions, which are present in large excess, have largely displaced the weakly bound TBA^+ ions.

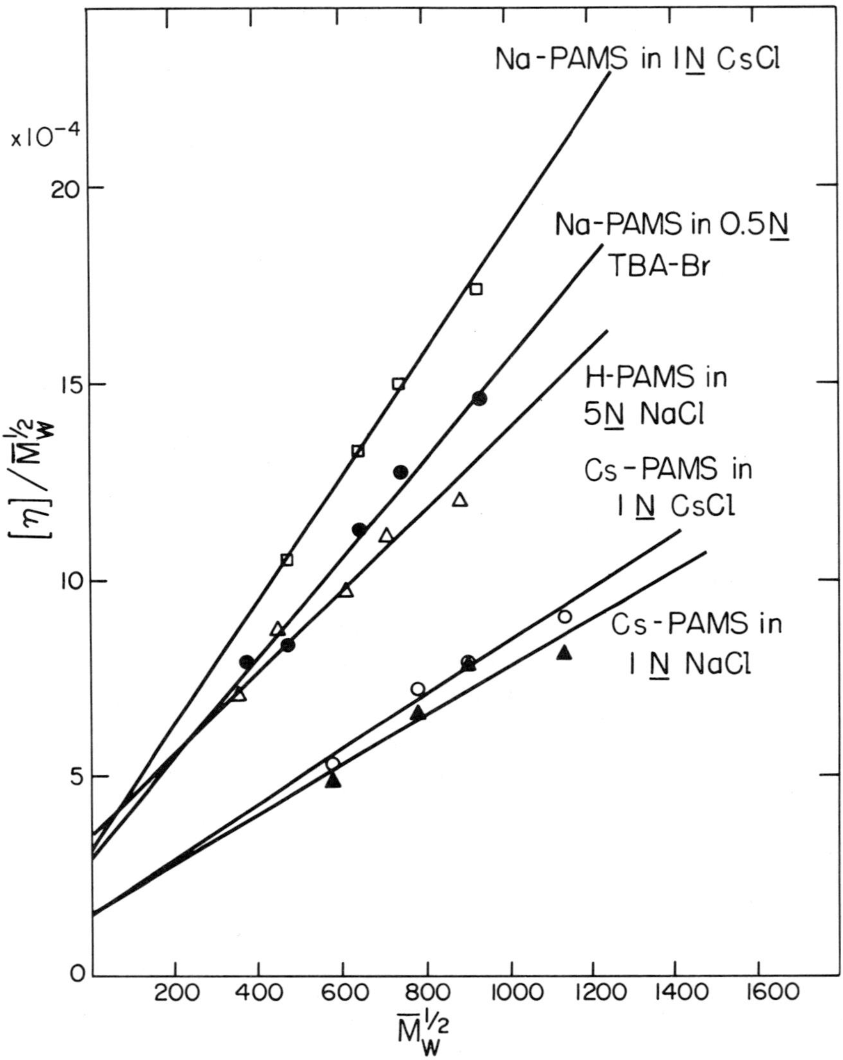

Fig. 3. The Stockmayer-Fixman plots for Na-PAMS, H-PAMS, and Cs-PAMS in various aqueous media.

Table II. Characteristic Ratios of PAMS Polymers

	H-PAMS	Na-PAMS			Cs-PAMS		TBA-PAMS		
	(NaCl)	(NaCl)	(CsCl)	(TBA-Br)	(NaCl)	(CsCl)	(NaCl)	(TBA-Br)	(organic)
$K_o \times 10^4$	3.5	3.0	3.2	3.0	1.6	1.6	1.0	3.0	2.0
C_∞	11.5	11.5	11.9	11.5	11.1	11.1	11.0	22.5	17.0

B. Evidence of Conformational Change by Fluorescence and Viscosity Data

1. *Characteristics of Fluorescence Probes*

A number of polycyclic aromatic compounds, usually containing either carboxylic or sulfonic acid groups to impart water solubility, have been found to be sensitive indicators of the polarity of their environment.[29,30] Light energy absorbed by molecules of this type may be emitted as fluorescent light or the excess energy may be dissipated by nonradiative transitions to the ground state. In highly polar media, where there is a strong interaction between excited molecules and their solvent shell, the latter mechanism is predominant. As the hydrophobicity of the environment increases, the number of radiative transitions increases, as shown in studies with organic solvents of varying polarity.[29] Hence, these probe molecules can be used to monitor the polarity of their environment.

In aqueous polymer solutions, the degree of fluorescence quenching is found to depend greatly upon the extent of clustering of hydrophobic groupings on the chains as determined by the ionic strength (C_s) of the medium.[22,31] In general, the nature and stoichiometry of the binding between probes and hydrophobic regions on macromolecules are poorly understood. No complete explanation has been demonstrated to account for the observed enhancement of fluorescence (an increase in the ratio of radiative to nonradiative transitions to the ground state) with an increase in hydrophobic density in some regions of the polymer domain. Partial shielding of the probe from solvent molecules by hydrophobic groupings is likely to be the major factor which can account for the observed phenomena. It has been found that the magnitude of the polymer-probe interaction has a large dependence on the specific groups involved.[29] Although lack of specific information makes it difficult to draw comparisons from fluorescence intensity data among polyelectrolytes containing different hydrophobic groups, it is clear that fluorescence quenching can be related to an increase in polarity of the environment of the probe and that in this respect the observations parallel the effects of solvent polarity.

2. *Results of Fluorescence Measurements for Sulfonate-Containing Homopolymers*

The fluorescence emission for all of the polyelectrolytes studied was found to be negligible in the absence of the TNS probe. Relative fluorescence intensities (F.I.) reported here represent the ratios between the emissions of the polymer samples and the emission of the quinine sulfate standard.

Relative fluorescence intensities for the homopolymers studied are plotted as a function of $C_s^{-1/2}$ in Figure 4A. The $C_s^{-1/2}$ dependence of this plot was chosen simply as a means of comparison with the $[\eta]$ data described below. A number of conclusions can be drawn after considering the shape of the curve for Na-PSS. It is apparent that the presence of phenylene groups provides a sufficient hydrophobic domain to cause a significant amount of fluorescence. The increase in hydrophobic clustering with ionic strength is indicated by the increase in F.I. The region of greatest slope in the curve is indicative of the C_s range in which the chain conformation is undergoing the greatest rearrangement. It may be viewed as a transition region between a highly extended and a compact polymer coil.[31] Finally, as might be expected, the change in conformation becomes much more gradual near the theta condition of the polymer (4.17 $\underline{N}$[26]). Na-PVS shows no fluorescence up to its theta condition (1.0 $\underline{N}$[24]). The absence of a hydrophobic side chain coupled with the high density of sulfonate groups probably eliminates the possibility of Na-PVS undergoing a conformational change of the type observed for Na-PSS. The Na-PMOS[32] and Na-PAMS[27] polymers are far from their theta condition in 5 $\underline{N}$ NaCl, and the fluorescence data indicate that Na-PAMS has not yet reached the region of radical conformational change at this point, while Na-PMOS is at the onset of this transition region. The difference in behavior between these two polymers may be due to the presence of α-methyl groups on the Na-PMOS backbone and to structural differences in the side chains which result in differences in polymer chain flexibility.

3. *Viscosity Results for Sulfonate-Containing Homopolymers*

The presence of conformational changes of which the fluorescent probe is a sensitive indicator should also be reflected in $[\eta]$ measurements on the polyelectrolytes in solutions of varying electrolyte concentration. The dependence of $[\eta]$ on C_s for polyelectrolytes has been treated[18,35-37] theoretically or semi-empirically in a number of cases.

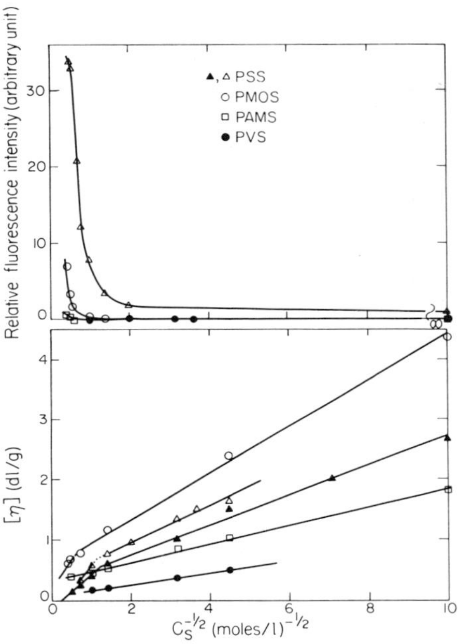

Fig. 4A. Relative fluorescence intensity vs. $C_s^{-1/2}$ for sulfonate-containing homopolymers in aqueous NaCl solutions.

Fig. 4B. [η] vs. $C_s^{-1/2}$ for the same polymers (Note: data for Na-PSS were taken from Ref. 14).

Takahashi and Nagasawa[18] showed that the parameter B of the Stockmayer-Fixman equation varied linearly with $C_s^{-1/2}$. Cox[36] derived limits for the dependence of $C_s^{-0.4}$ to $C_s^{-0.6}$ for freely draining and nondraining coils, respectively. It was reasonable for purposes of the present investigation to assume the value of -0.5 for the exponent of C_s.

Intrinsic viscosity is plotted against $C_s^{-1/2}$ for the various homopolymers in Figure 4B. Also included in Figure 4B are data obtained by Takahashi, Kato, and Nagasawa[26] for Na-PSS ($\bar{M}_w$ = 390,000) in NaCl. As shown in the graph, a linear relationship between [η] and $C_s^{-1/2}$ does not hold for all polyelectrolytes. The data for Na-PSS have been approximated by two intersecting straight lines. This treatment is supported by the more extensive data collected by Takahashi, Kato, and Nagasawa. Although a theoretical basis for the observed behavior is lacking, the intersection of the two straight lines occurs at the onset of the region of maximum change in conformation (as observed in the fluorescence experiment). A similar intersection has been assumed for the Na-PMOS data, although within experimental error these data could be represented by a single straight line. Unfortunately, the limit of NaCl solubility prohibits collection of further data which might substantiate a discontinuity. The behavior of Na-PVS and Na-PAMS is well represented for each by a single straight line. This is as expected since fluorescence data for these two polyelectrolytes show no indication of a conformational change. (The probable transition region for Na-PAMS has not been reached in the accessible ionic strength range.)

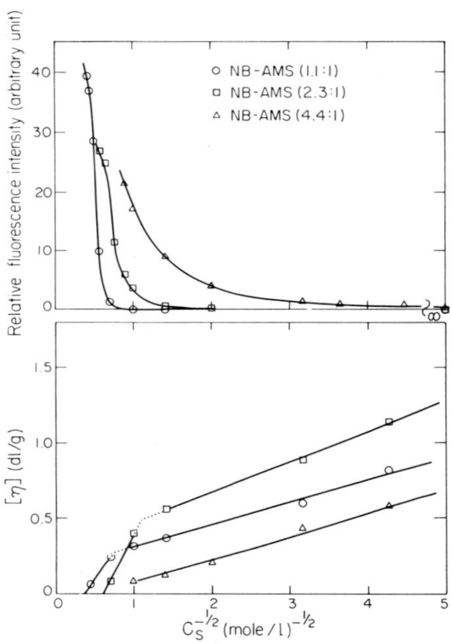

Fig. 5A. Relative fluorescence intensity vs. $C_s^{-1/2}$ for NB-AMS copolymers in aqueous NaCl solutions.

Fig. 5B. [η] vs. $C_s^{-1/2}$ for the same copolymers.

4. *Fluorescence and Viscosity Results for NB-AMS Copolymers*

In Figures 5A and 5B, F.I. and [η] are plotted as a function of $C_S^{-1/2}$ for the three NB-AMS copolymers. The solubility limits of these polymers, as roughly estimated by single cloud point determinations are as follows: Na-NB-AMS (1.1:1), 5.5 N LiCl; Na-NB-AMS (2.3:1), 3.5 N NaCl; and Na-NB-AMS (4.4:1), 2.0 N NaCl. It is observed that the transition region becomes better defined as the AMS content of the polymer increases. The Na-NB-AMS (4.4:1) curve does not exhibit the tendency to level off near the polymer precipitation point as is observed for the other samples. At any given ionic strength the magnitude of F.I. increases with increasing hydrophobic content of the copolymer, but greater intensity at the point of precipitation is observed for increasing AMS content. One might expect that a more radical change in conformation (a greater degree of hydrophobic clustering) would be required to arrive at a theta condition for the more hydrophilic polymers.

Figure 5B reveals that intersecting straight lines adequately describe the salt concentration dependence of intrinsic viscosity for the Na-NB-AMS (1.1 and 2.3:1) samples. In both cases, the points of intersection again fall at the onset of the conformational transition region indicated by the fluorescence measurements. It is assumed in the case of the Na-NB-AMS (4.4:1) polymer that the intrinsic viscosity data are not sensitive enough to give definite indication of the gradual conformational change observed.

IV. CONCLUSIONS

Intrinsic-viscosity data for several PAMS salts was interpreted by the Stockmayer-Fixman method in order to obtain the unperturbed dimensions of the polymer coils. The unperturbed dimensions of TBA-PAMS in aqueous TBA-Br were found to be larger than the corresponding value for TBA-PAMS in organic media. A still smaller value was calculated for PAMS with a variety of counterions in aqueous solutions containing small monovalent counterions.

Through the use of a fluorescence probe, conformational changes for a number of sulfonate-containing polymers were monitored as a function of ionic strength. Data obtained by this method shows a good correlation with $[\eta]$ measurements carried out on these polymers as a function of salt concentration. Na-PSS and NB-AMS copolymers exhibit conformational changes over various ranges of ionic strength, but no transition was observed for Na-PVS. The onset of a conformational transition occurs in Na-PAMS and Na-PMOS polymers near the limit of NaCl solubility.

REFERENCES

(1) Flory, P.J., "Statistical Mechanics of Chain Molecules," Interscience, New York, N.Y., 1969.

(2) Katchalsky, A., and Lifson, S., J. Polym. Sci. 11, 409 (1956).

(3) Kurata, M., and Stockmayer, W. H., Fortschr. Hochpolym.-Forsch. 3, 196 (1963).

(4) Bohdanecky, M., Polymer Letters 3, 201 (1965).
(5) Karagannidis, P., and Dondos, A., Makromol. Chem. 147, 135 (1970).
(6) Dondos, A., and Benoit, H., Macromolecules 4, 279 (1971).
(7) Tricot, M., Merteus, G., Collette, P., and Desreux, V., Bull. Soc. R. Sci., Liege 43, 502 (1974).
(8) Dondos, A., and Benoit, H., Eur. Polym. J. 4, 561 (1968).
(9) Dondos, A., and Benoit, H., Polymer Letters 7, 335 (1969).
(10) Dondos, A., Rempp, P., and Benoit, H., J. Polym. Sci. Part C 30, 9 (1970).
(11) Dondos, A., and Benoit, H., Macromolecules 6, 242 (1973).
(12) Sarkar, D.K., and Palit, S.R., J. Polym. Sci., Part C 30, 69 (1970).
(13) Abe, M., and Fujita, H., J. Phys. Chem. 69, 3263 (1965).
(14) Bereza, S.V., Bekturov, Ye. A., and Rafikov, S.R., Vysokomol. Soedin A11, 1681 (1969).
(15) Bekturov, Ye. A., and Legkunets, R. Ye., Vysokomol. Soedin A12, 626 (1970).
(16) Frisman, E.V., Sibileva, M.A., Ngan, N.T.K., and Nekrasova, T.N., Vysokomol. Soedin A10, (8), 1834 (1968).
(17) Flory, P.J., and Osterheld, J.E., J. Phys. Chem. 58, 653 (1954).
(18) Takahashi, A., and Nagasawa, M., J. Am. Chem. Soc. 86, 543 (1964).
(19) Wiederhorn, N.M., and Brown, A.R., J. Polym. Sci. 8, 652 (1952).
(20) Silberberg, A., Elliassaf, J., and Katchalsky, A., J. Polym. Sci. 23, 253 (1957).
(21) Nekrasova, T.N., Ptitsyn, O.B., and Shikanova, M.S., Vysokomol. Soedin A10, 1530 (1968).
(22) Tan, J.S., and Gasper, S.P., J. Polym. Sci., Polym. Phys. Ed. 12, 1785 (1974).
(23) Morawetz, H., Isr. J. Chem. 11, 173 (1973).
(24) Eisenberg, H., and Woodside, D., J. Chem. Phys. 36, 1844 (1962).
(25) Raziel, A., and Eisenberg, H., Isr. J. Chem. 11, 183 (1973).
(26) Takahashi, A., Kato, T., and Nagasawa, M., J. Phys. Chem. 71, 2001 (1967).
(27) Tan, J.S., and Marcus, P.R., J. Polym. Sci., Polym. Phys. Ed. 14, 239 (1976).
(28) Fisher, L.W., Sochor, A., and Tan, J.S., manuscript in preparation, to be submitted to Macromolecules.
(29) Edelman, G.M., and McClure, W.O., Accts. Chem. Res. 1, 65 (1968).
(30) Stryer, L., Science 162, 526 (1968).

(31) Tan, J.S., and Schneider, R.L., J. Phys. Chem. 79, 1380 (1975).
(32) Tan, J.S., and Gasper, S.P., J. Polym. Sci., Polym. Phys. Ed. 13, 1705 (1975).
(33) Eisenberg, H., and Mohan, G. Ram, J. Phys. Chem. 63, 671 (1959).
(34) Stockmayer, W.H., and Fixman, M., J. Polym. Sci., Part C 1, 137 (1963).
(35) Pals, D.T.F., and Hermans, J.J., Rec. Travs. Chim. 71, 433 (1952).
(36) Cox, R.A., J. Polym. Sci. 47, 441 (1960).
(37) Oosawa, F., "Polyelectrolytes," Marcel Dekker, Inc., New York, N.Y., 1971.

THE CHANGE OF STRUCTURE AND VISCOSITY OF STYRENE-ACRYLATE COPOLYMERS IN AQUEOUS MEDIA

O. C. C. Lin
E. I. duPont de Nemours & Company

I. INTRODUCTION:

The conformation of polymer electrolytes is dependent on the electrostatic interactions among the charge carrying groups. Such polymers can be shown to behave like coils of varying degree of extension and viscosity by changing the ionic characteristics of the medium.[1,2] Although the viscosity of polymer electrolytes in dilute solution and in dispersion has been extensively studied,[3,4] their behaviors in concentrated systems is yet not well understood. It has been shown, nevertheless, that overlapping molecular domains and molecular interactions existed in concentrated polymer solutions have given rise to certain effects not seen in dilute systems.[5,6] Further complications arise for copolymers consisting of groups of different polarity. In such systems hydrophilic or hydrophobic interaction of the constituent monomers is demonstratably an additional factor affecting the conformation, and consequently, the viscosity[7] The structure and viscosity of a concentrated copolymer system with different types of co-monomers can conceivably be quite complicated. Indeed, anomalous viscosity with water dilution has been shown for a number of polymers.[8,9,10,11] We shall report in this communication the peculiar viscosity behaviors of some polymer electrolytes consisting of acrylic acid, styrene and various acrylates in a wide range of concentration. In view of the fact that water based systems have received increased attention in recent years for its environmental impacts,[12] the viscosity behavior may be a significant parameter in the consideration of the processing and application characteristics of these systems.

II. MATERIALS AND METHODS:

All the styrene-acrylate copolymers used in this study were synthesized by free radical polymerization using peroxides as initiator. The molecular weights were relatively low, ranging from 7,000 to 10,000 (Mw). The Mw/Mn ratio was maintained from 2 to 3 and the monomer distribution on the backbone was basically random. The initial solids concentrations were from 68 to 70% in ethylene glycol monobutyl ether, unless otherwise specified. Approximately 40% of the -COOH was reacted with a long chain epoxides containing 12 to 14 carbon atoms to provide a constant ratio of -OH/-COOH. Introduction of the epoxides in the side chain has plasticizer action in the end use of the polymer as surface coating. It also increases the solubility or dilutability of the system with water. Its effect on the viscosity behavior of systems reported in this communication represents a constant factor throughout since the level used and the method of incorporation were kept constant in all the preparations. After neutralization to the desired level with diethyl ethanol amine, the samples were diluted with distilled water to various solids level for rheological measurements. No other inorganic salts, surfactants, or ionic species is presented in the system. Relevant data of the sample used are shown in the legend of the figures presented.

Most viscosity measurements were performed at 25°C and 8.5 sec^{-1} using either the Haake Rotovisko or the DuPont Gelometer. Both are concentric cylindrical rotational viscometers. Where viscoelastic measurements over a range of shear rates is required, the Weissenberg Rheogoniometer, Model R 18, was used.

III. RESULTS AND DISCUSSION:

A. The Viscosity as a Function of Dilution:

As water is added to the polymer solution, the viscosity decreases slowly and approaches a minimum followed by a remarkable increase with further addition of water. Eventually it reaches a peak which, in many cases, is higher than that of the initial polymer solution. As more water is added, the viscosity falls precipitously. After this point, the system displays behavior similar to an emulsion. The change of viscosity with dilution of a typical styrene-acrylate copolymer is shown in Figure 1. It is noted that over the narrow range of solids concentration around the peak, namely from 40% to 30%, the viscosity changes about 10,000 fold.

Concurrent with the viscosity change, the system shows many transitions of colloidal states. The initial solution, indicated by A in Figure 1 is clear and transparent. With dilution, it remains clear and transparent until it reaches the viscosity minimum, B where cloudiness starts to appear. The cloudiness increases as the system becomes more viscous,

C, and is essentially opaque at the viscosity peak, D. As dilution continues, E and F, the system becomes milky and fluid. Figure 2A shows photographs of the materials at these states. Figure 2B shows photographs taken immediately after they are turned sideway.

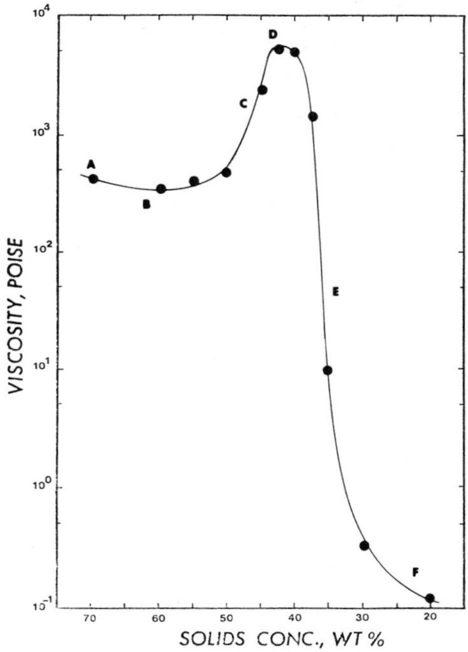

Fig. 1. The change of viscosity as a function of dilution. The viscosity is expressed as poise and the concentration as percentage of the polymer solids in the system. The material is a copolymer of styrene/acrylic acid (4/1), acid number 60 and with 100% neutralization.

In the region of the viscosity peak, the system is a gel. Under steady state shear at constant shear rate and constant temperature, the shear stress decreases with the time of shear and upon removal of the shearing action, the stress recovers with time toward its initial value. This thixotropic behavior is shown schematically in Figure 3 where f_o is the maximum stress displayed under the particular shear rate of the

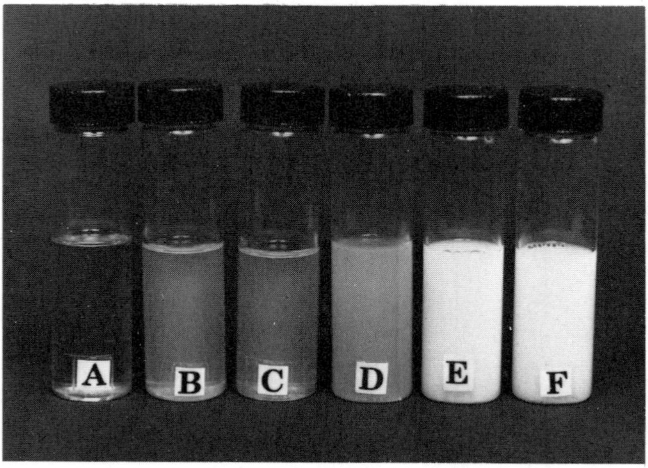

Fig. 2A. A photograph showing stages A to F as designated in Fig. 1. The tubes are in the up-right position. It is noticeable the A is transparent, B and C are cloudy, D is opaque, and E and F are milky white.

Fig. 2B. Same as Fig. 2A(above). The photograph was taken shortly after the tubes were turned sideway. The surface of the material in the tube provides a visual indication of the viscosity of the material.

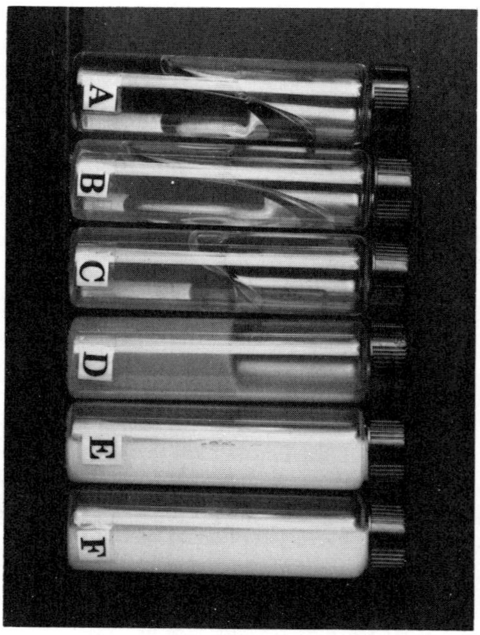

measurement and f_∞ is the asymptotic stress attained at equilibrium:

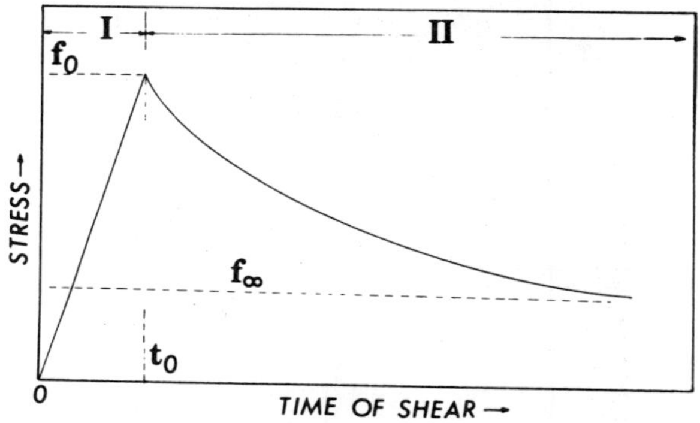

Fig. 3. A schematic diagram showing the change of shear stress as a function of the time of shearing, under constant shear rate and temperature. This figure defines the phenomenon known as thixotropy. (See Reference 13 for detailed discussions).

The coefficient of thixotropy defined according to the following equation:(13)

$$\lambda = \frac{f_0 - f_\infty}{f_\infty}$$

was measured. It varied from 0.5 to 5 depending on the shear rate and the shear history. Thus all the rheological measurements made at this state are highly sensitive to the condition of shear. Figure 4 are actual curves showing the change of the shear stress as a function of the time of shearing, under a constant shear rate 2.28 sec^{-1} and at 25 C,of an identical sample but with different shear history.

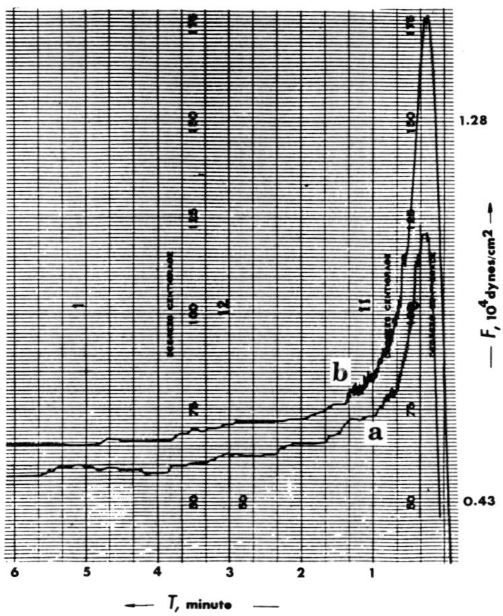

Fig. 4. Typical strip-chart recordings of thixotropy. The material is that of stage D designated in Fig. 1. Curve a is for the material that has been rested for 30 minutes in the viscometer before measurement at 2.28 sec.$^{-1}$ Curve b is that for a 60 minutes rest period. See Reference 13 for detailed experimental procedures of thixotropy.

The solution characteristics of the system at stages of A and B is further demonstrated by a constant low shear viscosity, (Figure 5, two upper curves). At states E and F, the viscosity is essentially Newtonian, (Figure 5, lower curves).

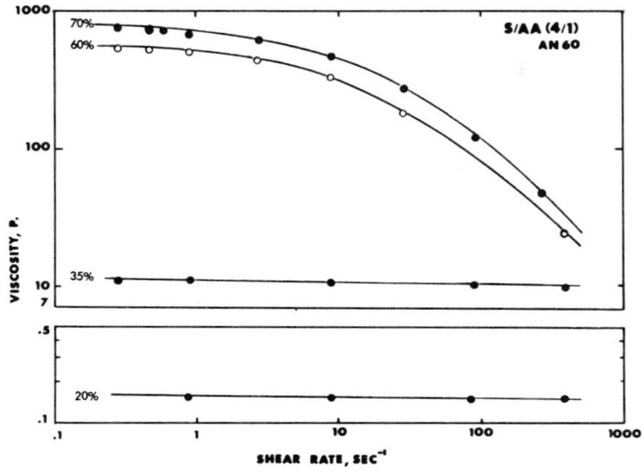

Fig. 5. The change of viscosity as a function of the shear rate - shear thinning effect. The material is same as in Fig. 1., namely, a co-polymer of styrene/acrylic acid (4/1), acid number 60 with 100% neutralization.

The molecular interactions leading to the gel network is conceivably similar to that leading to micelle structure.[14] Basically, the systems studied consist of molecules of long hydrophobic chain (polymer backbone) with hydrophilic,charged groups distributed randomly throughout. In a sense, they are polymeric surfactants. The physical chemistry of such systems has been quite extensively studied in the recent years, notably by Graham and Holden.[15] In view of the solids concentration level studied, which is normally 10^2 times that of the usual surfactant systems, and the viscosity encountered, which is 10^3 - 10^4 times higher, the term "micelle" or"micelle formation" may be misleading for these systems. It appears more appropriate rather to characterize the system in the realm of gel.

B. Effects of the Degree of Neutralization

The viscosity of polymer electrolyte solutions has been shown to depend strongly on the degree of ionization of the electrolytes.(16,17,18) In the copolymer systems studied, the degree of ionization is approximated by the degree of neutralization as controlled by stoichiometric addition of amine according to the level of acrylic acid. As a proton is removed or bound by the neutralizing base, a negatively charge center is created on the backbone which interacts with other negative centers along the backbone leading to extended chain conformation and consequently, higher viscosity. Shown in Figure 6 are the viscosity with water dilution of the same polymer as in Figure 1, (Styrene/Acrylic Acid, 4/1), neutralized to various degree from 50% to 120%.

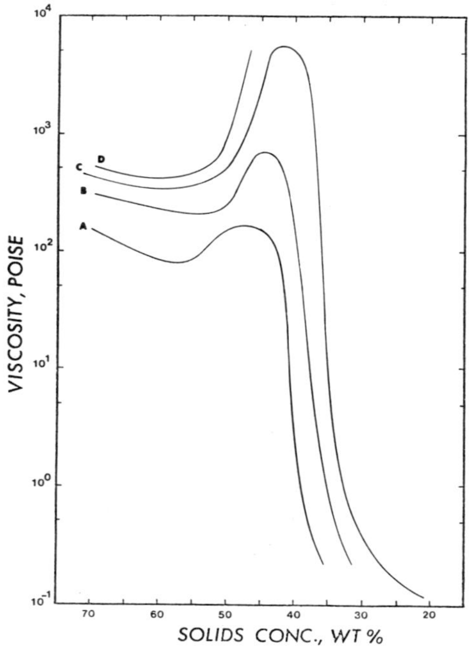

Fig. 6. The effects of the degree of neutralization on the viscosity with water dilution. The material is same as in Fig. 1, namely, styrene/acrylic acid (4/1), acid number 60. The degree of neutralization is for A, 50%; B, 70%; C, 100% and D, 120%.

It is noticed that at the same solids level, the viscosity increases with the degree of neutralization as expected. At stages around the peak, increased neutralization leads to a marked increase in viscosity, over orders of magnitude in the 120% case. This enormous viscosity increase is actually much too large to be accounted for by the expansion of polymer chains alone. The simultaneous development of gel structure suggests that other factors are contributing. For example, the amine used for neutralization is likely to promote intermolecular entanglements through better solubilization of the polymer similar to the case of polyesters.(10) Simultaneously as the medium becomes more polar, interactions between non-polar groups are enhanced. Both these phenomena will result in the formation of network. The relative importance on viscosity of chain extension due to electrolytic charges and of interaction due to solvency balance was independently studied by Hill and Brandenburger (11), who concluded that the latter was the dominant factor. At the emulsion states, the viscosity is also affected by the degree of neutralization, but to a lesser extent. At the lower than 50% neutralization, the system is unstable leading to separation of two phases, as expected from the basic concept of Katchalsky. (19)

C. Effects of Co-Monomer Composition:

To examine the effect of hydrophobic clustering, changes in co-monomer composition is in order. Whereas ionizable carboxyl groups are necessary for a system to be dilutable with water, the actual dilution curve is found dependent on the composition of the co-monomers. Curve A in Figure 7 represents the dilution viscosity of a copolymer consisting of styrene and acrylic acid which is similar to the copolymer used for Figure 1 but at a lower acid number. When diluted with water, the system display transition of states and viscosity behavior similar to those shown in Figure 1. The initial viscosity decrease is largely due to dilution of the polymers. As more water is added, the ionized carboxylates start to show effects in extending the chain conformation. Simultaneously the non-polar phenyl groups cluster in the medium as this is becoming increasingly polar. Both these factors hasten the formation of intermolecular networks leading to higher and higher viscosity. As part of the styrene is replaced by butyl acrylate, which is less hydrophobic and has more hydrogen-bonding capability than styrene, the viscosity decreases (Curve B). This is presumably due to fewer phenyl groups for hydrophobic clustering and increased flexibility of the molecules in adjusting to the new medium. Replacement of the styrene by methyl methacrylate, which is more hydrophilic than either butyl acrylate or styrene, further changes the viscosity behavior. Indeed the last system shows a viscosity behavior similar to that of a normal polymer solution, the viscosity anomaly is only slight, (Curve C). The reduced clustering of the hydrophobic groups is aided by the increased H-bonding capability of the polymer when styrene is replaced by butyl acrylate and by methyl methacrylate. Increased polymer/solvent interaction makespolymer/polymer clustering less favorable.

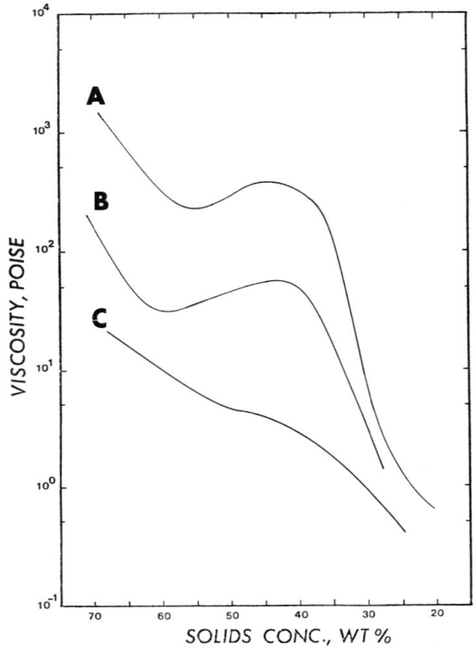

Fig. 7. The effect of co-monomer composition on the viscosity with water dilution. The monomer composition for A is styrene/acrylic acid (3.5/1); B is styrene/butyl acrylate/acrylic acid (2/1.5/1); and C is methyl methacrylate/butyl acrylate/acrylic acid (2/1.5/1). All have the same acid number, 50 and with 70% neutralization.

D. Effects of Solvent

The viscosity behavior, especially that of the dispersion states, is strongly affected by the composition of the solvent in the original polymer solution. This is shown in Figure 8. It is noted that the polymers in all cases have similar solids concentrations, molecular weights and were neutralized to 70% by diethyl ethanol amine.

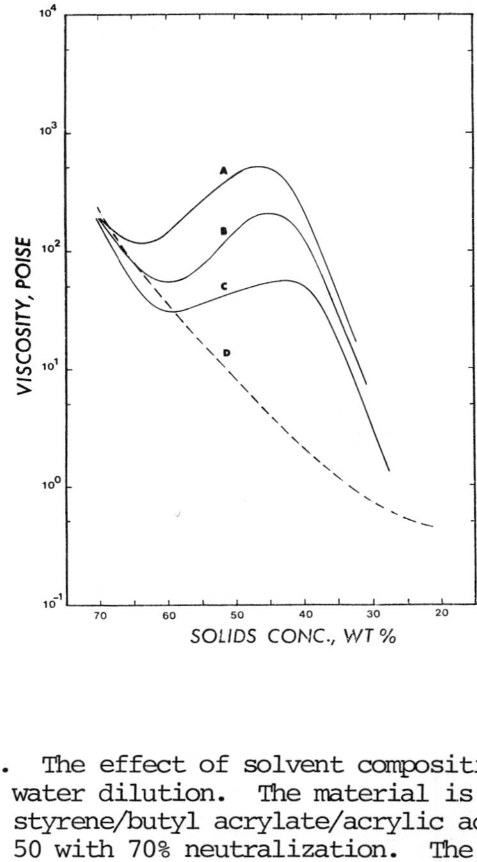

Fig. 8. The effect of solvent composition on the viscosity with water dilution. The material is same as in Fig. 7B, namely, styrene/butyl acrylate/acrylic acid (2/1.5/1), acid number 50 with 70% neutralization. The solvent is for A, ethylene glycol mono-n-butyl ether/ethylene glycol mono methyl ether (1/1); for B,ethylene glycol mono-n-butyl ether/ isopropanol (1/1); and C, ethylene glycol mono-n-butyl ether. Curve D is for dilution of the material by ethylene glycol mono-n-butyl ether, which is the original solvent of the polymer.

The successive increase in viscosity especially around the peak of these systems is unmistakable. A tabulation of the solubility parameters of the solvents are shown below:

Solvent	Solubility Parameter By Hansen's Method (20,21) Total	Non-Polar	Polar	H-Bonding	Association
Ethylene glycol mono-methyl ether	12.1	7.9	4.5	8.0	9.2
Isopropanol	11.5	7.7	3.0	8.0	8.5
Ethylene glycol monobutyl ether	10.2	7.8	2.5	6.0	6.5

The last column is the "association interaction" parameter, δ_a, defined as:

$$\delta_a = (\delta_p^2 + \delta_H^2)^{\frac{1}{2}}$$

representing the combined effects of polar interaction, δ_p, and H-bonding, δ_H.(22) A qualitative correlation of and the viscosity behavior as shown in Figure 8 is apparent. It appears that solvents having the larger polar and H-bonding parameters give the higher viscosity. This is in agreement with premise that clustering of hydrophobic groups and non-polar moiety become more severe in a polar medium.

E. Reversibility of the Viscosity/Dilution Behavior

The unusual viscosity behavior with dilution is reversible. The same viscosity/concentration relationship as shown in Figure 1 can be obtained if one adds varying amounts of the concentrated polymer solution (Stage A) to the latex emulsion (Stage F) and mix thoroughly. Of more practical interest with regards to their application, for example as surface coating, is the reversibility of the relationship as water and solvent evaporates. To examine this aspect, the attenuated reflection of pulsues (ARP) method discussed by Myers (23) was applied. The system in the diluted stage is applied as a thin film on the surface of a fused quartz crystal. A shear wave of ultrasonic frequency (10×10^6 c.p.s) is directed at the coating and substrate interface. The change in mechanical impedance, due to solidification of the film, is calculated by measuring the attenuation and phase change of the reflecting wave. Figure 9 shows the dynamic viscosity calculated from such measurements on a coating consisting of a solution polymer (Curve A) and a water reducible polymer (Curve B), during the solidification process.[24]

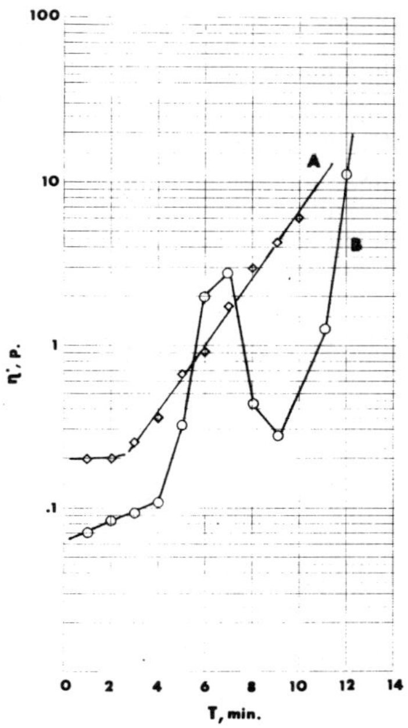

Fig. 9 The change of the dynamic viscosity as a function of solvent evaporation. Dynamic viscosity is calculated from measurements in the ultrasonic impedometer, (see Reference 23 and the references there-in). Curve A is for poly methyl methacrylate in a mixed solvent of acetone and toluene. Curve B is for the polymer system described in Fig. 6, curve B, namely styrene/butyl acrylate/acrylic acid (2/1.5/1).

In the former the dynamic viscosity increases monotonously with evaporation. In the latter phase transitions occur resulting in a peak in the dynamic viscosity. It should be pointed out that the viscosity/evaporation curve (Curve B) is not expected to be an exact reversal of the dilution curve for two reasons. First, the evaporation measurement was done under very high frequency (cyclic shear) whereas the dilution measurement was done under relatively low shear rate (steady state shear). Second, both the compositions of water and solvent are continuously changing during the evaporation process. The qualitative similarity between the evaporation and dilution curves is apparent, however. The position (in composition) and the level of viscosity where the peak occurs are important variables of the system. It is also noted that the time of evaporation corresponding to the dynamic viscosity peak is in agreement with a model recently suggested (25) for the evaporation of water and solvent from coating systems,

IV. SUMMARY AND CONCLUSIONS:

Acrylic and styrene copolymers containing partially neutralized carboxylic acids display unusual structural changes and viscosity behavior when diluted with water. In successive dilutions the viscosity decreases initially, reaches a minimum, increases to a peak and falls precipitously afterward. This behavior is associated with the transitions of physical chemical states - from solution to dispersion, to gel, to emulsion. The changes of viscosity in the solution state is due to dilution and depends on the degree of neutralization as expected from the theories of polymer electrolytes. The dispersion is believed to result from increased clustering of hydrophobic groups such as styrene which counteracts forces for conformation extension due to electrostatic repulsion of the ionized carboxyl groups. These interactions become more severe as more water is added leading to the formation of gel structures. The rheological behavior of the gelled network is shear dependent and thixotropic. With further addition of water, the network collapses resulting in a latex emulsion. These changes can be qualita tively portrayed by a model shown in Figure 10. The linkage bridging the particles can be depicted schematically as shown in Figure 11. The general feature of this model is indicated by the observation that, for a given type of co-monomer composition, the change of viscosity is strengthened by polar solvents capable of forming H-bonding and that, for a given solvent, strengthened by hydrophobic monomers. Between stages D and E, conversions of inter-particle linkages to particles occur leading to an abrupt collapse of the gel structure and viscosity.

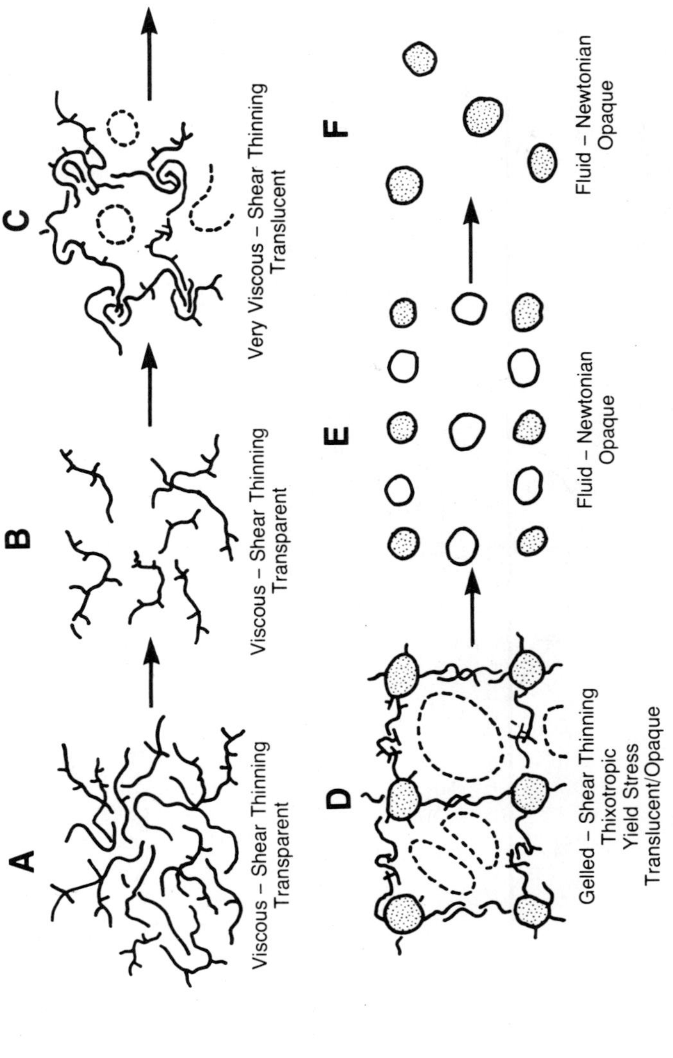

Fig. 10 A schematic diagram portraying the changes in molecular arrangements occuring in water dilutable polymers during successive additions of water. See text for explanation.

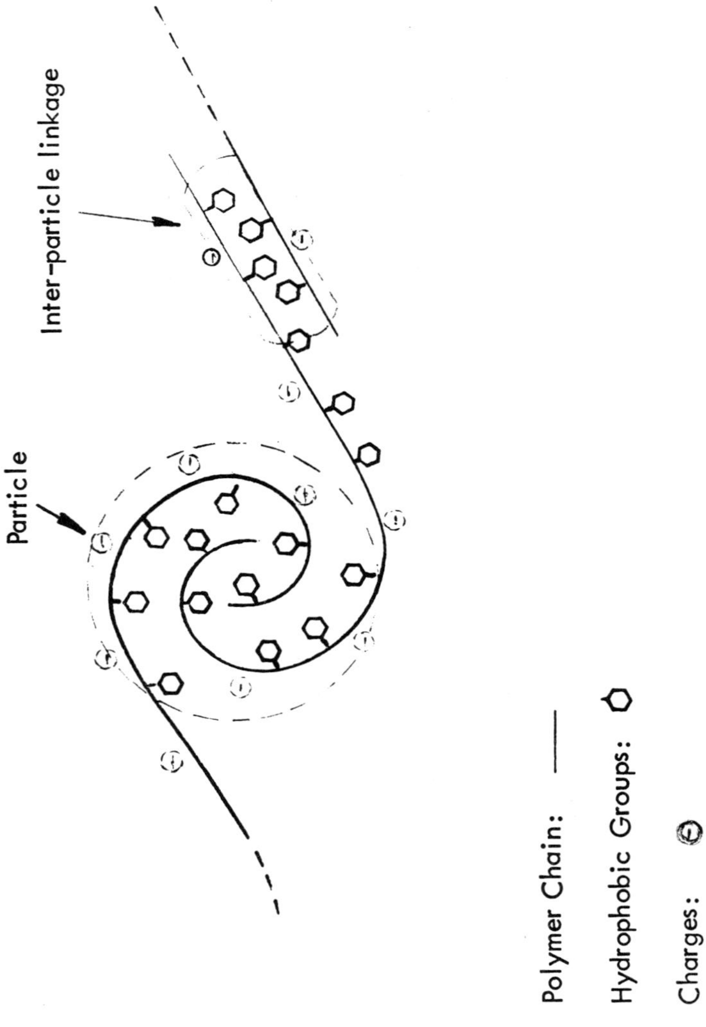

Fig. 11 A schematic diagram portraying a possible version of the structure and interaction inside the particles and in the interparticle linkages.

ACKNOWLEDGEMENTS

I would like to thank many of my colleagues who have made invaluable contributions to this work especially: W. Van Hoeven, R. N. Iacona, C. W. Secker, I. Hazan, L. F. Nonemaker, A. G. Armour and W. S. Zimmt. I would also like to thank Mrs. Betty Giordano for assistance in preparing this manuscript and finally, E. I. duPont de Nemours & Company for an opportunity to work in the subject area and for the approval to publish this paper.

REFERENCES:

1. Doty, P. and Ehrlich, G. Ann. Rev. Phys. Chem. 3, 81 (1952).
2. Pols, D.T.F. and Hermans, J.J., Rec. tran. chem., 71 433 (1952)
3. Tanford, C., "Physical Chemistry of Macromolecules," John-Wiley and Sons, 1961, p. 457 ff.
4. Krieger, I.M., Adv. Colloid Interf. Sci., 3, (2) 111 (1972).
5. Berry, G.C., and Fox, T. G., Adv. Polymer Sci., 5, 261-357, (1968)
6. Gandhi, K.S. and Williams, M.C., J. Polym. Sci. part C, No. 35, 211 (1971).
7. Ferry, J.D., Udy, D.C., Wu, F.C. Heckler, G.E. and Fordyce, D.B., J. Colloid Sci., 6, 429 (1951)
8. Benerji, G. and Bretts, R.A., British Paint Res. Assoc. Res. Memo. No. 359, (1965).
9. McEwen, I.H., J. Paint Tech., 45 (583),33 (1973)
10. Doerffel, J. Farbe + Lack, 81 (1) 10 (1975).
11. Hill, L.W. and Brandenburger, L.B., Progress Org. Coating, (1976). In press.
12. Bikales, N.M., Editor, "Water Soluble Polymers," Plenum Press, New York, 1973.
13. Lin, O. C. C., J. Appl. Polym. Sci., 19, 199 (1975)
14. Adamson, A.W., "Physical Chemistry of Surfaces," 2nd Edition Interscience, 1967, p. 473 ff.
15. Graham N.B. and Holden H.W., Polymer, 10 (8), 633 (1969)
16. Markovitz H., andKimball, G.E., J. Colloid Sci., 5, 115 (1950).
17. Oth, A., and Doty, P., J. Am. Chem. Soc. 56, 43 (1952).
18. Nada, I., Tsuge, T. and Nagasawa, M., J. Phy. Chem., 74 No. 4, 710, (1970).
19. Katchalsky, A., J. Polymer Sci., 7, No. 4, 393 (1952).
20. Hansen, C.M., J. Paint Tech., 39, 511 (1967)
21. Burrell, H., in "Polymer Handbook," J. Brandrup and E. H. Immergut, Editors, 2nd Edition, Interscience, New York, 1975.

22. Barton, A.F.M., Chem. Rev., 75 (6), 731 (1975.
23. Myers, R.R., J. Polym. Sci., Part C., No. 35, 1 (1971).
24. Lin, O.C.C., Presentation at 46th Annual Meeting, Society of Rheology, St. Louis, Mo., October, 1975.
25. Sullivan, D.A., J. Paint Tech., 47 (610), 60 (1975).

STRUCTURE-SOLUBILITY RELATIONSHIPS IN POLYIMIDES

Frank W. Harris and Lynn H. Lanier
Wright State University

I. INTRODUCTION

A number of methods have been used to attain solubility in polyimides while maintaining their desirable high temperature characteristics. The majority of approaches have involved the following structural modifications: (1) the incorporation of flexible or non-symmetrical, thermally-stable linkages in the backbone, (2) the introduction of large polar or non-polar substituents along the polymer backbone, and (3) the disruption of symmetry and recurrence regularity through the copolymerization of two dianhydrides or two diamines. Although broad generalities associated with structural effects on solubility are impossible, interesting structure-solubility comparisons can be made within series of closely related polyimides. Tables 2-10 have been derived from various literature sources to illustrate some of these structure-solubility relationships.

The polyimides which have been reviewed are limited to polymers that do not undergo thermal decomposition in air below 350°, as indicated by thermogravimetric analysis. Polymers containing more than one aliphatic carbon in the backbone repeat unit are not included unless the carbons are perfluorinated. This review also does not include polyamide-imides, polyester-imides, or polyimides containing other heterocyclics in the backbone. The majority of the polyimides reported were synthesized by polymerizing a dianhydride with a diamine.

TABLE 1

Solvent and Dianhydride Abbreviations

Solvent or Dianhydride	Abbreviations
Acetone	A
Benzene	B
m-Cresol	C
Chloroform	Cf
Dioxane	D
Dimethylformamide	DMF
Dimethylacetamide	DMAC

TABLE 1 (cont'd)

Solvent or Dianhydride	Abbreviations
Dimethylsulfoxide	DMSO
Nitrobenzene	NB
N-Methyl-2-pyrrolidone	NMP
Methylene Chloride	MC
Pyridine	P
Tricresol Mixture	TC
sym-Tetrachloroethane	TCE
Trifluoroacetic Anhydride	TFA
DMF and DMAC	Amide
3,3',4,4'-Benzophenonetetracarboxylic Acid Dianhydride	BTDA
3,3',4,4'-Diphenylethertetracarboxylic Acid Dianhydride	ETDA
Pyromellitic Dianhydride	PMDA

II. SOLUBLE POLYIMIDES CONTAINING FLEXIBLE OR NON-SYMMETRICAL LINKAGES

A. Perfluoroalkyl Linkages

A series of polyimides which were solubilized by incorporating perfluoroalkyl links in the polymer backbone are shown in Table 2. Incorporating perfluoroisopropyl groups in the diamine did not result in a PMDA-based soluble polyimide (1). However, when the perfluoroisopropyl was also included in the dianhydride a soluble polymer was obtained (2). This polymer is offered commercially by E. I. duPont de Nemours Company.

Perfluoromethylene groups can also be used in the backbone to attain solubility (3-6). However, at least three perfluoromethylene groups were required to solubilize polymer 3. Solubility was not attained with a BTDA-based polyimide even when five perfluoromethylene groups were included in the diamine (7). Soluble BTDA-based polyimides, however, were obtained when perfluoroalkyl ether links were included in the diamine (8 and 9). Soluble polymers were also prepared from these diamines and dianhydrides containing perfluoroalkyl and perfluoroalkyl ether linkages (10 and 11). In the series of polyimides 8 through 11, the solubility of the polyimides increased as the number of flexible linkages in the repeat unit increased. The increased flexibility also resulted in lower glass transition temperatures.

TABLE 2

Soluble Polyimides Containing Perfluoroalkyl Linkages

Polymer Ref	Dianhydride	Diamine	Solvents
		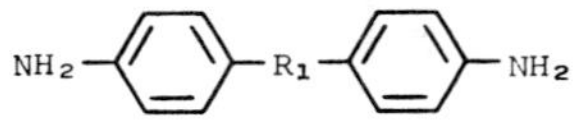	
1 [1]	PMDA	$R_1 = -C(CF_3)_2$	insol.
	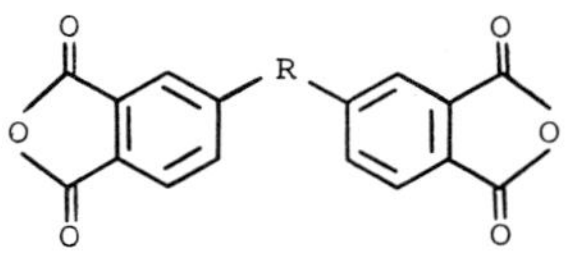		
2 [2]	$R = -C(CF_3)_2$	$R_1 = -C(CF_3)_2$	A,B,Cf,D
3 [3,4]	$R = (CF_2)_3$	$R_1 = -O-, -CH_2-, -SO_2-$	DMAC
		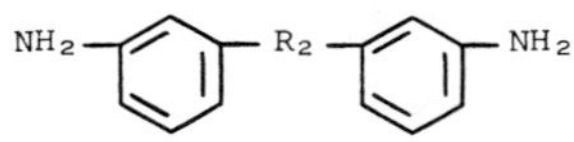	
4 [3,4]	$R = (CF_2)_3$	$R_2 = -O-$	DMAC
5 [3]	$R = (CF_2)_3$	$R_2 = (CF_2)_3$	DMF
6 [3]	ETDA	$R_2 = (CF_2)_3$	DMF
7 [3]	BTDA	$R_2 = (CF_2)_5$	insol.
8 [5]	BTDA	$\left\{ \begin{array}{l} -O(CF_2CF(CF_3)O)_x(CF_2)_5 \\ (OCF(CF_3)CF_2)_yO- \end{array} \right\} x+y=2$	DMAC, A
9 [5]	BTDA	$\left\{ \begin{array}{l} -O(CF_2CF(CF_3)O)_x(CF_2)_5 \\ (OCF(CF_3)CF_2)_yO- \end{array} \right\} x+y=3$	DMAC, A
10 [5]	$R = (CF_2)_6$	$\left\{ \begin{array}{l} -O(CF_2CF(CF_3)O)_x(CF_2)_5 \\ (OCF(CF_3)CF_2)_yO- \end{array} \right\} x+y=3$	DMAC, A
11 [5]	$R = -O(CF_2)_5O-$	$\left\{ \begin{array}{l} -O(CF_2CF(CF_3)O)_x(CF_2)_5 \\ (OCF(CF_3)CF_2)_yO- \end{array} \right\} x+y=3$	DMAC, A

B. Oxygen and/or Sulfur Linkages

A series of soluble polyimides containing oxygen and/or sulfur linkages are shown in Table 3. It appears the polyimides based on PMDA cannot be solubilized by merely incorporating ether linkages in the diamine component (12). Solubility can be attained, however, by employing several ether links in both the dianhydride and the diamine (14, 16). Multiple ether linkages are evidently needed as the incorporation of a single ether linkage in the dianhydride and the diamine results in a polymer that is only soluble at low molecular

TABLE 3

Soluble Polyimides Containing Oxygen or Sulfur Linkages

Polymer[Ref]	Dianhydride	Diamine	Solvents
		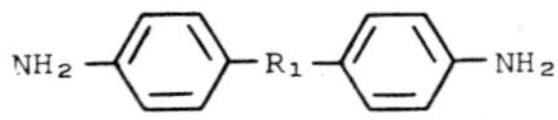	
12[6]	PMDA	R_1 = 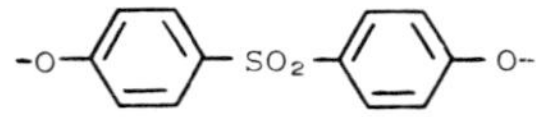	insol.
13[6]	BTDA	R_1 = 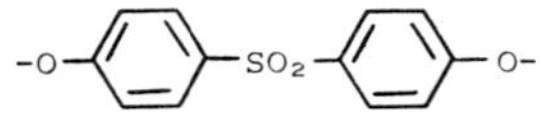	DMAC
	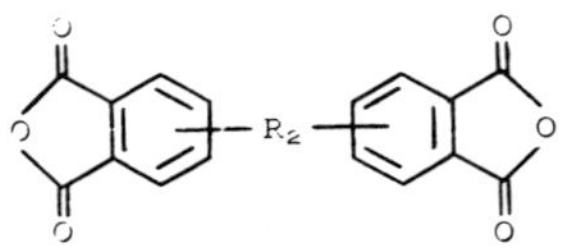	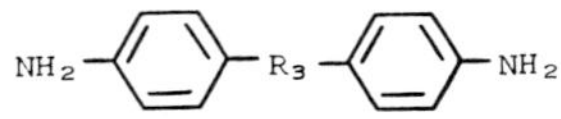	DMAC
14[7]	R_2 = *o*, 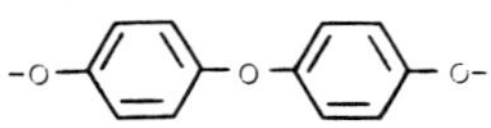	R_3 = -O-	Cf
15[8]	R_2 = *m*, 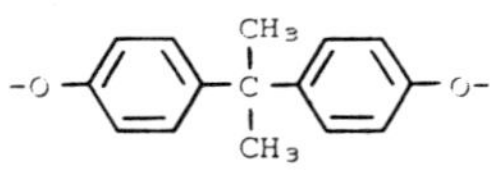	R_3 = -O-	insol.[a]
16[8]	R_2 = *m*, 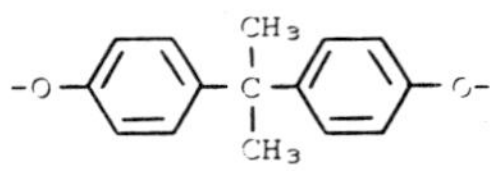	R_3 = -O-	DMF[b]
17[9]	R_2 = *m*, $-SO_2-$	R_3 = -C(=O)-	NMP
18[10]	ETDA	R_3 = -O-	Formal

(a) The polymer was heat treated at 275° to complete imidization.
(b) The polymer was imidized in refluxing o-dichlorobenzene.
(c) The polymer was soluble at low molecular weight in 10 parts phenol:7 parts 2,4,6-trichlorophenol.

weights (18). A soluble polyimide can also be prepared from BTDA using a diamine containing two ether linkages and a sulfonyl linkage (13).

Soluble polyimides have been prepared from 3,3',4,4'-diphenylsulfonetetracarboxylic acid dianhydride (17). The sulfonyl group, however, tended to encourage side reactions leading to low molecular weight polymers.[11,12]

C. Phosphorous Linkages

Several soluble polyimides have been prepared from a dianhydride containing a phosphorous linkage (Table 4). Although the reaction of the dianhydride with 4,4'-diaminodiphenylmethane

TABLE 4

Soluble Polyimides Containing Phosphorous Linkages

Polymer[Ref]	Dianhydride	Diamine	Solvents
		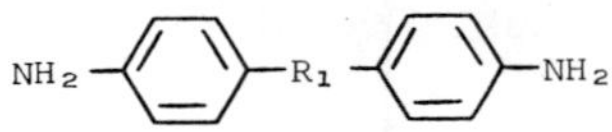	
19[13]	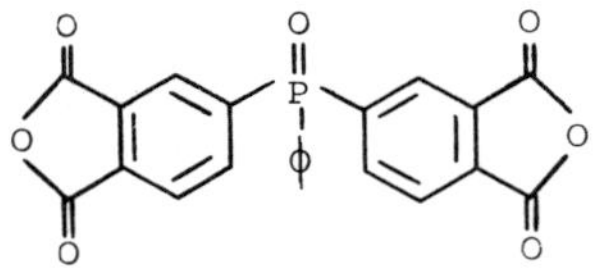	R_1 = $-CH_2-$	insol.
20[13]	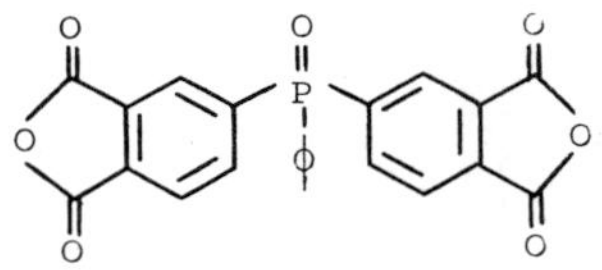	R_1 = -O- 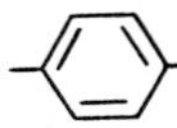 -O-	insol.
21[13]	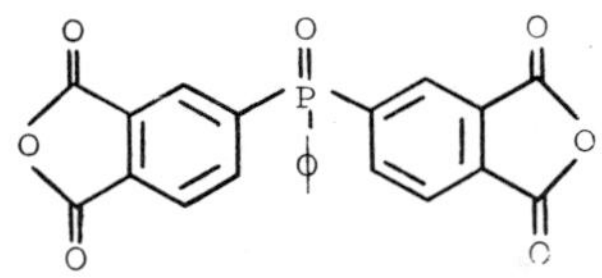	R_1 = -O-, -S-	DMAC, DMF, NB
22[14]	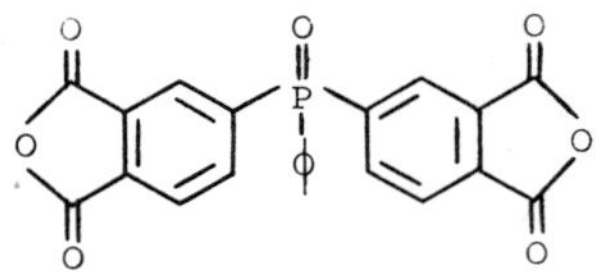	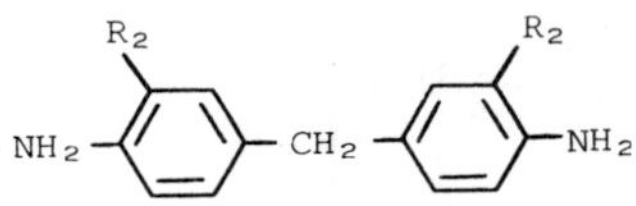 R_2 = $-CH_3$ or $-OCH_3$	(a)

(a) The solvents were not reported, however, it was stated to have improved solubility over polymer 19.

gave an insoluble polymer (19), the polymerization of the dimethyl- and dimethoxy-substituted derivatives of this diamine afforded soluble polyimides (22). Interestingly, the use of a diamine containing two ether linkages resulted in the insoluble polymer 20, while the polymerization of a diamine containing a single ether linkage gave a soluble product (21). All of the polymers containing phosphorous exhibited outstanding thermal stability.

III. SOLUBLE POLYIMIDES CONTAINING LARGE PENDANT SUBSTITUENTS

A. Soluble Polyimides Containing Non-Polar Substituents

Several examples of soluble polyimides containing non-polar pendant substituents are shown in Table 5. In addition

TABLE 5

Soluble Polyimides Containing Non-Polar Substituents

Polymer[Ref]	Dianhydride	Diamine	Solvents
23[10]	PMDA	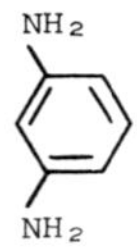	insol.
24[15,16]	PMDA	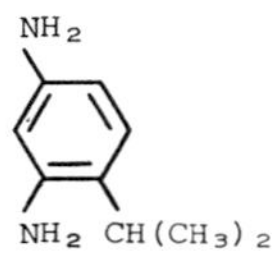	DMF
25[17]	PMDA, BTDA	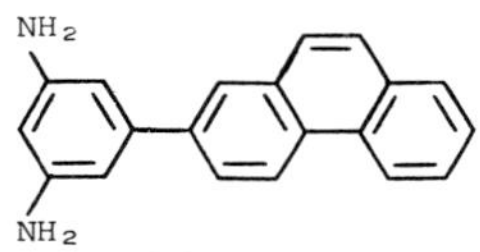	DMF, DMAC, DMSO, NB, NMP
26[18]	PMDA	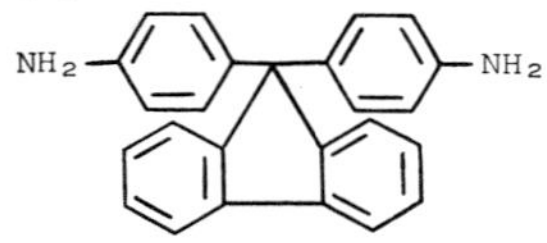	insol
27[18]	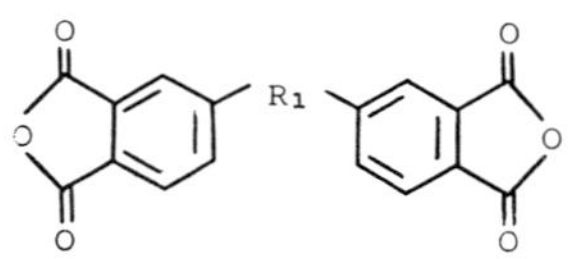 R_1 = nil, $-SO_2-$, $-\mathrm{C}(=\mathrm{O})-$, -O-	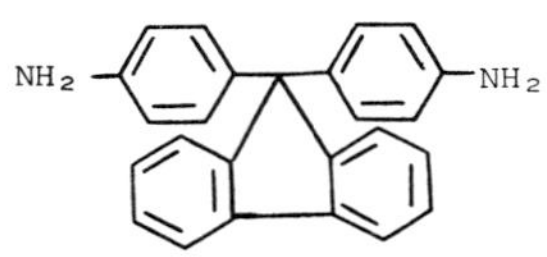	TCE, DMF
28[19,20]	BTDA	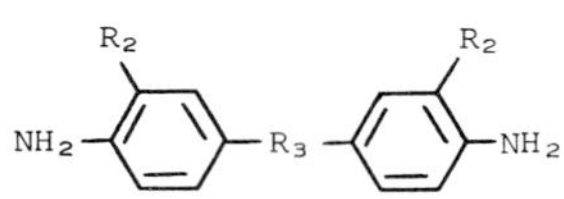 R_2 = -H, $-CH_3$; R_3 = $-CH_2-$	C[a]

(a) These polymers were imidized by heating at 120° to 180°.

to the substituents, all of the polymers also contain a non-symmetrical or flexible linkage in the repeat unit. Hence, it is likely that the introduction of pendant, non-polar groups along a polyimide backbone does not in itself result in solubility.

In the polymer series 27 an evaluation of the various linkage's relative abilities to aid in solubility was carried out. The number of solvents in which the resulting polymer was soluble increased in the following order:[18]

$$\text{nil} \approx -SO_2- < -\underset{\underset{O}{\|}}{C}- < -O-$$

A series of phenylated polyimides has been prepared in this laboratory (Table 6).[21,22] The polymers were obtained by the polymerization of phenylated dianhydrides with aromatic diamines in *m*-cresol containing isoquinoline. The polymerization mixtures were stirred at ambient temperature for 1 hr and then heated at 202-203° for 3 hr. Under these conditions, the intermediate polyamic acid was converted to the corresponding polyimide. Interestingly, polymer 29a, which does not contain a flexible linkage, was the only polymer that precipitated from the *m*-cresol-isoquinoline mixture. All of the polymers are soluble in chlorinated hydrocarbons.

B. Polar Substituents

Several soluble polyimides that contain polar pendant substituents have been prepared (Table 7). As in the previous series, it appears that a non-symmetrical or flexible linkage is also needed to attain solubility.

The effects of polar and non-polar substituents on solubility are very similar. For example, polymer 30 which contains sulfonamide substituents and polymer 24 (Table 5) which contains isopropyl groups are both soluble in DMF. Polar substituents, however, do tend to afford solubility in a greater number of solvents.

Polar substituents also vary in their ability to affect solubility. Polymer 34 which contains phenylamine substituents is soluble in TFA while the corresponding polymer containing acetamide groups (33) is insoluble. This is probably due to their difference in basicity. In the polymer series 37, polymers prepared with pendant phthalide groups are soluble in more solvents than those prepared from anilinephthalein. This has been attributed to the former's ability to undergo hydrogen bonding with protic solvents.

TABLE 6

Soluble, Phenylated Polyimides

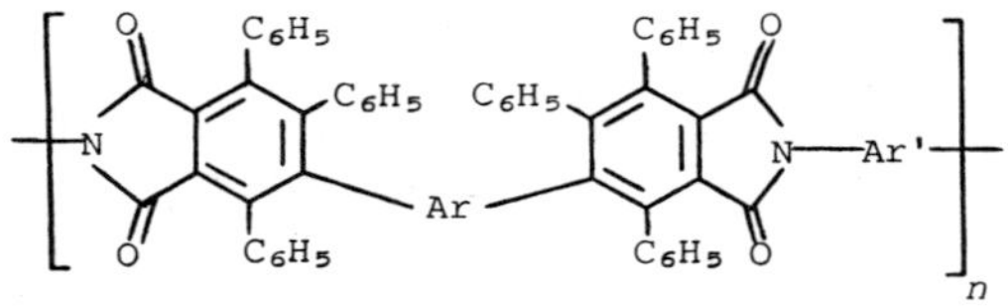

Polymer	Ar	Ar'	$[\eta]^{a}$	T_g^{b}
29a		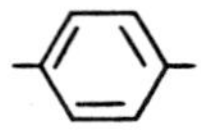	0.35	466°
29b		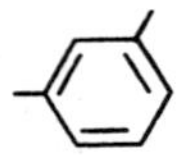	0.86	416
29c		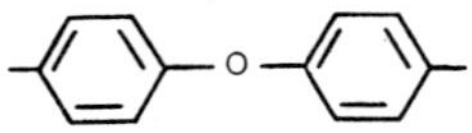	1.11	413
29d		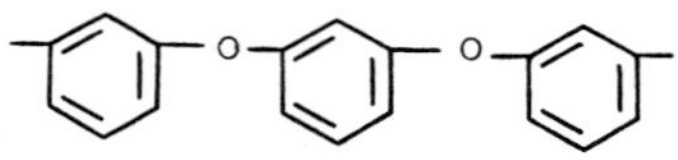	1.22	280
29e	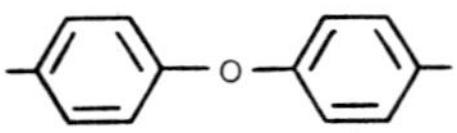	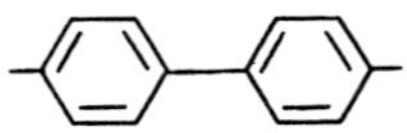	4.03	399
29f		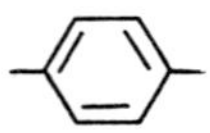	0.40	371
29g		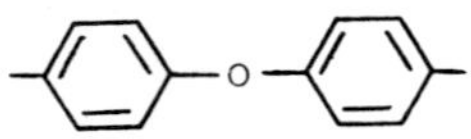	2.80	360
29h		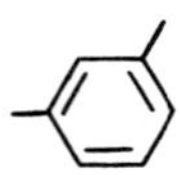	1.06	357
29i		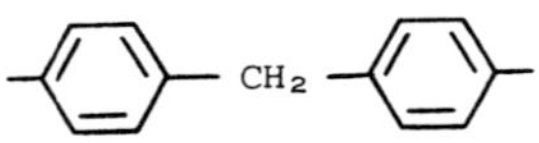	1.38	356
29j		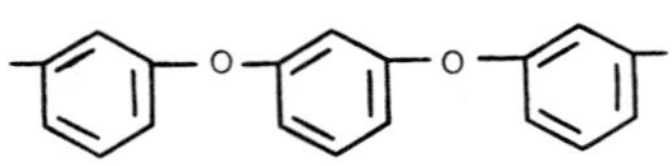	0.75	261

(a) Intrinsic viscosity obtained at 30° in *sym*-tetrachloroethane.

(b) Average values of the highest penetration rate on TMA at ΔT=20°/min (10 g load).

TABLE 7

Soluble Polyimides Containing Polar Substituents

Polymer[Ref]	Dianhydride	Diamine		Solvents
30[23]	PMDA	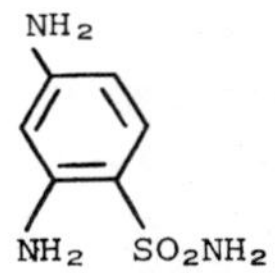		DMSO, DMF, NMP
31[11]	PMDA	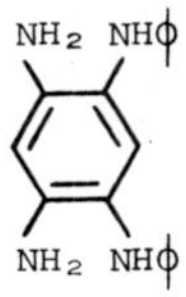		TFA[a]
32[11]	BTDA, ETDA	"		TFA, DMSO, DMAC, DMF
		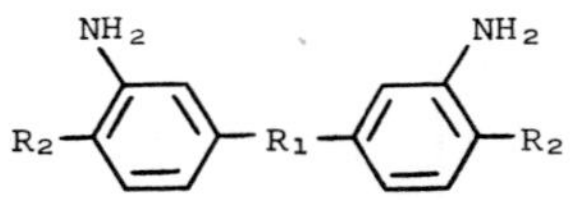		
		R_1 =	R_2 =	
33[12]	PMDA	nil	$-NHCOCH_3$	insol.
34[11]	PMDA	nil	$-NHC_6H_4$	TFA[a]
35[12]	PMDA, BTDA, ETDA	-O-, $-CH_2-$, $-OC_6H_4O-$	$-NHCOCH_3$	DMF, DMSO, DMAC, NB
36[11]	BTDA, ETDA	nil	$-NHC_6H_4$	DMF, DMAC, DMSO, NB
37[18]	PMDA, BTDA, ETDA	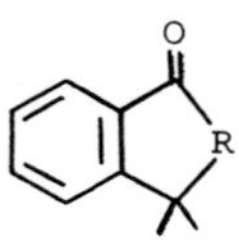 R = -NH-, -O-	-H	DMF, DMAC, NB, TC
38[24]	PMDA, BTDA, ETDA	-O-, $-CH_2-$, -C(=O)-, $-OC_6H_4O-$	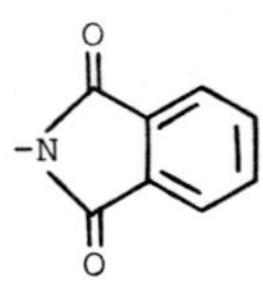	Polar Solvents
39[25]	PMDA	-O-	-OH	DMAC, DMF, DMSO

(a) The polymer was insoluble in DMF, DMSO, DMAC, and NB.

IV. SOLUBLE COPOLYIMIDES

Soluble polyimides have been prepared by copolymerizing two different dianhydrides with an aromatic diamine (Table 8). Similarly, the polymerization of a dianhydride with two diamines or two diisocyanates has afforded soluble polyimides (Table 9). The ratio of the two comonomers required to attain solubility depends on the monomers employed. For example, incorporating 50% of a phosphorous containing dianhydride into a PMDA-based polyimide failed to produce a soluble polymer (41). The use of 80% of the phosphorous containing monomer, however, gave the soluble polymer 40 (Table 8). Conversely, in a study with perfluoroalkylene linked polyimides, the incorporation of only 10% PMDA into a previously soluble system resulted in an insoluble polymer (46).

TABLE 8

Soluble Copolyimides From Two Dianhydrides

Polymer[Ref]	Dianhydride A	Dianhydride B	Diamine	Solvents
	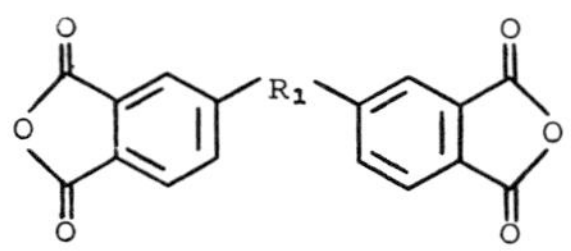		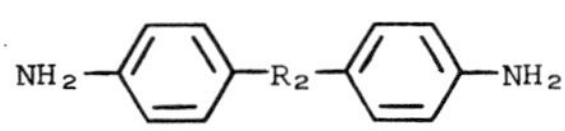	
40[13]	$R_1 = -P(=O)(C_6H_5)-$, 80%	PMDA 20%	R_2 = -O-	Amide
41[13]	R_1 = " , 50%	" 50%	R_2 = "	Insol.
42[26]	R_1 = nil, 50%	" 50%	R_2 = 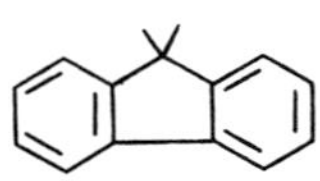	Amide, TCE
43[26]	R_1 = " , 100%	" 0%	R_2 = "	TCE
44[26]	R_1 = " , 0%	" 100%	R_2 = "	Insol.
45[26]	BTDA	ETDA	R_2 = "	Cf, MC, Amide, NB[a]
46[3]	$R_1 = (CF_2)_3$, 90%	PMDA, 10%	$R_2 = -CH_2-$	Insol.

(a) After thermal treatment at 300° the copolymer became insoluble in NB.

TABLE 9

Soluble Copolyimides From Two Diamines or Two Diisocyanates

Polymer[Ref]	Dianhydride	Amine A	Amine B	Solvents
		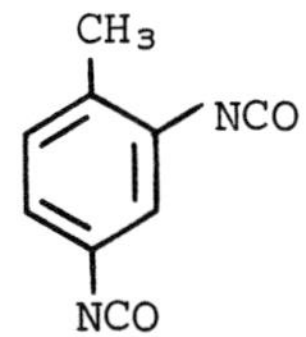	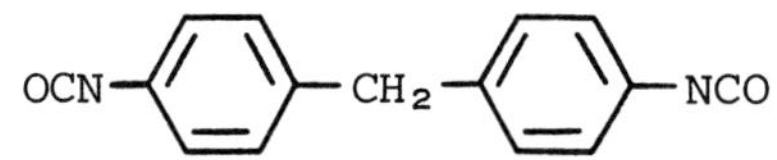	
47[27-29]	PMDA	80%	20%	NMP
48[27-29]	PMDA	100%	0%	Insol.
49[30]	PMDA	0%	100%	Insol.
50[25]	PMDA	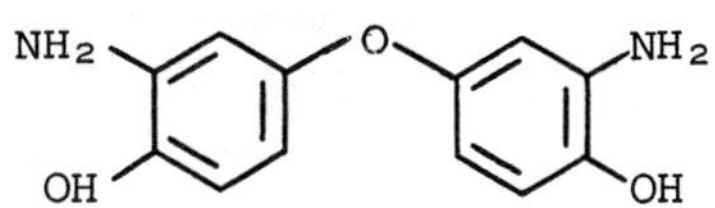	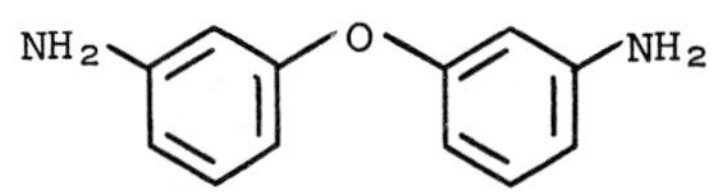	DMAC, DMF, NMP, DMSO

An excellent example of a beneficial effect of copolymerization on solubility is illustrated by polymers 42, 43, and 44. Polymer 43, which was prepared from 100% diphenyldicarboxylic acid dianhydride, is only soluble in TCE. The 100% PMDA-based polymer 44 is completely insoluble. The copolyimide 42 prepared from 50% of each of these dianhydrides is soluble in amide solvents as well as TCE.

Another example of a favorable copolymerization effect is the soluble polymer 47 resulting from the polymerization of toluene diisocyanate and 4,4'-methylenebis(phenyl isocyanate) with PMDA (Table 9). In this case, the polymers prepared from PMDA and each diisocyanate alone are completely insoluble (48 and 49). Copolymer 47 is offered commercially by The Upjohn Company.

V. EFFECT OF IMIDIZATION CONDITIONS ON SOLUBILITY

The extent to which crosslinking occurs during the preparation of aromatic polyimides is still controversial. It has been suggested that the initial polyimides were insoluble due to crosslinking.[31,32] Crosslinking was thought to occur when the soluble polyamic acid precursors were thermally dehydrated. Recent studies,[33-35] including the syntheses of soluble polyimides, however, suggest that polyimides are not generally crosslinked.

Whether crosslinking occurs or not, the imidization conditions employed in the formation of a polyimide can affect its solubility. For example, polymer 51 (Table 10) can be dehydrated at 185° to afford a soluble material. Heating the polymer at 300°, however, results in an insoluble resin. The insolubility may be due to crosslinking or simply to complete ring closure. The BTDA used to prepare this polymer is very susceptible to crosslinking through the reaction of its carbonyl group with primary amines.

The polyimide (53) prepared from BTDA and anilinefluorene is soluble when chemically imidized. Thermal imidization of this polymer at 300°, however, affords an insoluble material. The fact that a similar polymer prepared from diphenyldicarboxylic acid dianhydride and anilinefluorene is soluble when thermally imidized suggests that the carbonyl group is involved in crosslinking at the high temperature.

A perfluoromethylene-linked polyimide (4, Table 2) that is readily soluble in polar aprotic solvents has been prepared by a single-stage, melt polymerization technique. The synthesis of this polymer via the two-stage polyamic acid route produced a material that was only slightly soluble in DMAC after warming and stirring for three days. Evidently, substantial crosslinking occurred when the isolated intermediate was thermally dehydrated.

TABLE 10

The Effect of Imidization Conditions on Solubility of Polyimide Polymers

Polymer[Ref]	Dianhydride	Diamine	Cyclization	Solvents
51[36]	BTDA	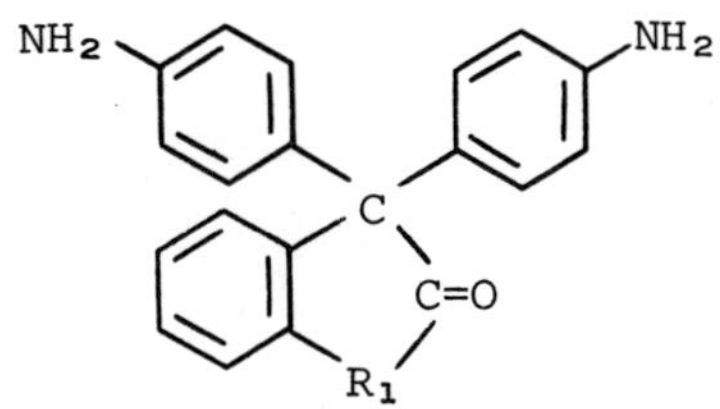 R_1 = -S-, -NH-	185°C	NMP
52[36]	BTDA		300°C	Insol.
53[26,37]	BTDA	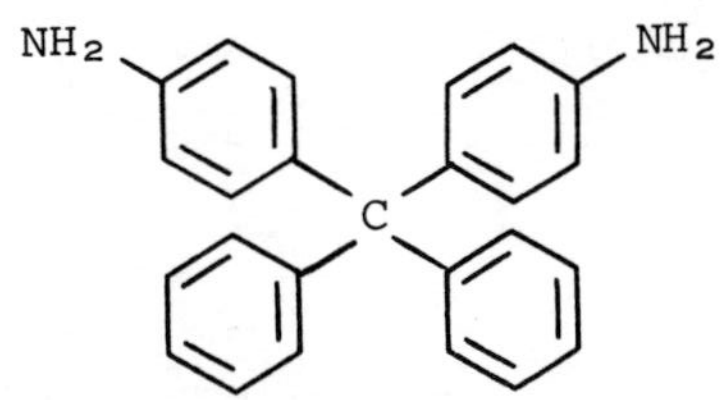	chemical	Amide
54[26]	BTDA		300°C	Insol.

VI. EFFECT OF CRYSTALLINITY AND INTERMOLECULAR BONDING ON SOLUBILITY

The exact role that crystallinity plays on polyimide solubility is difficult to define. The great majority of polyimides that are crystalline are insoluble. Several polyimides which contain perfluoromethylene linkages, however, are crystalline and soluble in DMF, e.g., polymer 5 (Table 2). Of course, there are many examples of completely amorphous polyimides that are insoluble.

Whether crystallinity is formally developed or not, there is evidently considerable intermolecular bonding in all polyimides due to the presence of the polar carbonyl groups. In fact, there is evidence that pyromellitimide units tend to form strong intermolecular bonds consistent with the presence of charge transfer complexes.[38] Apparently, separating the anhydride functions decreases these interactions. The results of two studies show that solubility is much more sensitive to the incorporation of flexible linkages in the dianhydride component than in the diamine.[26,28]

VII. ACKNOWLEDGMENT

The support of a portion of the phenylated polyimide work by the Air Force Materials Laboratory, Air Force Systems Command, United States Air Force, Wright-Patterson AFB, Ohio 45433, is gratefully acknowledged.

VIII. REFERENCES

1. Livshits, B. R., Vinogradova, S. V., Knunyants, I. L., Berestneva, G. L., and Dymshits, T. Kh., Vysokomol. Soedin., A, 15: No. 5, 969-968, (1973). Polym. Sci. USSR A 15, 1078-1087 (1973).
2. Rogers, F. E., U. S. Patent 3,356,648 (1967).
3. Critchley, J. P., Grattan, P. A., White, M. A., and Pippett, J. S., J. Polym. Sci., A-1, 10, 1789 (1972).
4. Critchley, J. P., and White, M. A., J. Polym. Sci., 10, 1809 (1972).
5. Webster, J. A., Butler, J. M., and Morrow, T. J. "Advances in Chemistry Series No. 129," Am. Chem. Soc., 1973, p. 61.
6. Brode, G. L., Kwiatkowski, G. T., and Hawakami, J. H., Am. Chem. Soc., Polymer Preprints 15 (1), 761 (1974).
7. Takekoshi, T., and Kochanowski, J. E., U. S. Patent 3,833,546 (1974); Chem. Abstr., 82, 4948f (1975).

8. Heath, D. R., and Wirth, J. G., U. S. Patent 3,875,116 (1975); Chem. Abstr., 84, 45207y (1976); Chem. Abstr., 83, 59838w (1975).
9. Badische, Anilin- and Fabrik, Soda-, French Patent 2,162,411 (1973).
10. Cotter, R. J., and Matzner, M., "Ring Forming Polymerizations, Heterocyclic Rings", Part B, 2, Academic Press, New York and London, 1972, pp. 4-126.
11. Korshak, V. V., Rusanov, A. L., Tugushi, D. S., Gitis, S. S., Ivanov, A. V., and Botkina, Y. V., Vysokomol. Soedin., A, 14 (3), 686 (1972).
12. Korshak, V.V., Rusanov, A. L., Katsarava, R. D., and Niyazi, F.F., Vysokomol. Soedin., A, 15 (12), 2643 (1973); Polym. Sci. USSR, 15, 3000 (1973).
13. Fedotova, O. Ya., Gorokhov, V. I., Paresishvili, O. I., Karetnikor, G. S., and Kolesnikov, Vysokomol. Soedin., A, 14 (6), 1256 (1972).
14. Fedotova, O. Ya., Popova, N. M., Fedotova, N., Vysokomol. Soedin., B, 15 (10), 760 (1973).
15. Angelo, R. J., Belgium Patent 655,654 (1965); Chem. Abstr., 65, 2376 (1966).
16. Wallach, M. L., Am. Chem. Soc., Polymer Preprints, 8 (2), 1170 (1967).
17. Stoenciu, A., Fedotova, O., Korshak, V. V., Tr. Mosk. Khim.-Tekhnol. Inst. 74, 130 (1973); Chem. Abstr., 82, 43766q (1975).
18. Vinogradova, S. V., Slonimskii, G. L., Vygodkiii, Ya. S., Askadskii, A. A., Mzhel'skii, A. I., Churochkina, N. A., and Korshak, V. V., Vysokomol. Soedin., A, 11 (12), 2725, (1969); Polym. Sci. USSR, 11, 3098 (1969).
19. Dietrich, W., German Patent 2,115,478 (1972); Chem. Abstr., 78, 17790r (1973).
20. Hosokawa, E., Waki, M., and Fukushima, M., Japanese Patent 74 14,959 (1970); Chem. Abstr., 82, 18202v (1975).
21. Harris, F. W., Feld, W. A., and Lanier, L. H., J. Polym. Sci., Polym. Let. Ed., 13, 283 (1975).
22. Harris, F. W., Feld, W. A., and Lanier, L. H., in "Applied Polymer Symposium" (N. Platzer, Ed.), No. 26, p. 421. John Wiley and Sons, New York, 1975.
23. Imai, Y. and Koga, H., J. Polym. Sci., A-1, 11, 2623 (1973).
24. Korshak, V. V., Rusanov, A. L., Katsarava, R. D., Niyazi, F. F., and Batirov, I., Vysokomol. Soedin., A, 16 (4), 722 (1974); Chem. Abstr., 81, 121095r (1974).
25. Korshak, V.V., Vinogradov, A. V., Al-Khaider, Ziyad, T., Tseitlin, G. M., and Rode, V. V., Vysokomol. Soedin., B, 14 (8), 592 (1972); Chem. Abstr., 78, 4613k (1973).
26. Vinogradova, S. V., Churochkina, N. A., Vygodskii, Ya. S., and Korshak, V. V., Vysokomol. Soedin., 15, 1713 (1973).
27. Farrissey, W. J. and Andrews, P. S., U.S. Patent 3,787,367 (1974).

28. Farrissey, W. J. and Andrews, P. S., U. S. Patent 3,870,677 (1975).
29. Farrissey, W. J. and Rausch, K. W., U. S. Patent 3,870,674 (1975).
30. Critchley, J. P., in _Progress in Polymer Science_, Vol. 2, (A. D. Jenkins, Ed.), Pergamon Press, Oxford, 1970, pp. 51-161.
31. Bernier, G. A., Kline, D. E., _J. Appl. Polym. Sci._, 12, 593 (1968).
32. Hermans, P. H. and Streef, J. N., _Makromol. Chem._, 74, 133 (1964).
33. Vinogradova, S. V., Korshak, V. V., and Vygodskii, Ya. S., _Vysokomol. Soedin._, 8, 809 (1966).
34. Korshak, V. V., Vinogradova, S. V., Vygodskii, Ya. S., and Yudin, B. N., _Izv. Akad. Nauk SSSR, Ser. Khim._, 6, 1405 (1968).
35. Korshak, V. V., Vinogradova, S. V., Vygodskii, Ya. S., and Buko, L. V., _Izv. Akad. Nauk SSSR, Ser. Khim._, 10, 2267 (1967).
36. Darmory, F. P. and DiBenedetto, M., U. S. Patent 3,705,869 (1972); _Chem. Abstr._, 78, 59024z (1973); _Chem. Abstr._, 79, 67293x (1973).
37. Korshak, V. V., Vinogradova, S. V., Vygodskii, Ya. S., Gerashcenko, N. V., and Lushkina, N. I., _Vysokomol. Soedin., A_, 14 (9), 1924 (1972); _Polym. Sci. USSR_, 14, 2153 (1972).
38. Dine-Hart, R. A., _Makromol. Chem._, 143, 189 (1971).

SOLUBILITY-STRUCTURE STUDY OF POLYIMIDES

T. L. St.Clair, A. K. St.Clair, and E. N. Smith
NASA-Langley Research Center

I. INTRODUCTION

Aromatic polyimides are a class of polymers with excellent thermal stability, but their utility has been somewhat limited by their insolubility in organic solvents (1,2). The primary approach in the past in obtaining solubility of these polymers has been the introduction of pendent aromatic groups along the backbone (3-5). The purpose of this investigation was to improve the solubility of aromatic condensation polyimides by a dual approach. First, novel dianhydride monomers with bridging groups intended to impart solubility were incorporated into the polymer chain; and second, the isomeric points of attachment of bridging groups in the diamine monomers were varied to further enhance solubility.

II. EXPERIMENTAL

A. Materials

4,4'-Oxydiphthalic anhydride (ODPA) (6) was recrystallized from anisole and sublimed at 200°C-210°C/1mm prior to use (m.p. 224°C). 2,2-Bis(3,4-dicarboxyphenyl)hexafluoropropane dianhydride (6F) (7) was purified by sublimation at 210°-220°C/1mm, m.p. 239°C. 2,2-Bis[4-(3,4-dicarboxyphenoxy)-phenyl]hexafluoropropane dianhydride (BFDA) (8), m.p. 217°C, and 4,4'-bis(3,4-dicarboxyphenoxy)diphenyl sulfide dianhydride (BDSDA) (9), m.p. 189°C-190°C, are research anhydrides and were used as received. The 3,3',4,4'-benzophenone tetracarboxylic dianhydride (BTDA) and pyromellitic dianhydride (PMDA) were obtained from commercial sources and purified by vacuum sublimation, m.p. 224°C and 285°C respectively. The diamine monomers were acquired from commercial sources or prepared and purified as previously reported by Bell and coworkers (10). The *m*-phenylenediamine (*m*-PDA) was purified by sublimation, m.p. 59°C and *p*-phenylenediamine (*p*-PDA) was purified by recrystallization from distilled water and sublimation, m.p. 138°C. The dimethylacetamide (DMAc) used as a solvent for polymerization was vacuum distilled at 102°C from calcium hydride.

B. Preparation of Polyamic Acids

Polymerizations were conducted at 15% solids (w/w) by adding the diamine (0.0015 mole) and DMAc to a flask flushed with dry nitrogen. The appropriate dianhydride (0.0015 mole) was then added as a solid in a single portion, and the solution was stirred at room temperature for 18-24 hours. The resulting polymer solutions were refrigerated until needed for casting.

C. Thermal Conversion of Polyamic Acids to Polyimides

The polyamic acid solutions were poured onto soda-lime glass plates in an enclosed dust-free area at a relative humidity of 10-20%. Solutions were spread with a 15 mil doctor blade to ensure a final film thickness of approximately 1 mil. A 4 mil blade gap was used to produce thinner film suitable for infrared analysis. Films of the polyamic acids were dried in air for 16 hours at room temperature, then thermally converted to the polyimides by heating in a forced air oven or in dry nitrogen 1 hour each at 100°, 200°, and 300°C.

D. Chemical Conversion of Polyamic Acid to Polyimide

To an almost colorless solution of ODPA-*o*,*p*'-ODA polyamic acid prepared by the method discussed above, was added a solution of acetic anhydride (.0045 mole) and pyridine (.0045 mole) in 1 ml DMAc. After being stirred for 8 hours at room temperature, the resulting dark yellow solution was poured into distilled water in a blender. The yellow fibrous polyimide that precipitated was washed several times with distilled water and dried *in vacuo* at 120°C for 6 hours.

E. Characterization

Melting points were determined on an E.I. duPont Model 900 Differential Thermal Analyzer at 10°C/min. Inherent viscosities of polyamic acid and polyimide solutions were determined at a concentration of 0.5% in DMAc at 35°C. Infrared spectra of polyimide films were obtained on a Perkin-Elmer Model 137 infrared spectrophotometer. Films heated to 300°C were tested for solutility at room temperature in N-methylpyrollidone (NMP), DMAc, N,N-dimethylformamide (DMF) or *m*-cresol. If a film was insoluble at the end of 3 days, the solution was gently heated in a warm water bath. BFDA polyimide films were also tested at room temperature for solubility in yellow fuming nitric acid at approximately 1% solids (w/w). Thermomechanical analyses (TMA) were performed on films in static air at a 5°C/min. temperature program (11). Torsional braid analyses (TBA) (12) were performed on glass fiber braids coated with polyamic acid solutions heated to 300°C in nitrogen at 3°C/min. The procedure used for thermogravimetric analysis (TGA) has been reported elsewhere (13).

III. RESULTS AND DISCUSSION

A. Polymerization

Shown in Figures 1 and 2 are the structures for the dianhydrides and aromatic diamines used during this investigation. The reaction of ODPA, 6F, BFDA, BDSDA, BTDA and PMDA with isomeric variations of diaminodiphenylmethane (MDA), diaminobenzophenone (DABP), diaminodiphenyl ether (ODA) and phenylenediamine (PDA) yielded, for the most part, high molecular weight film-forming polymers. Inherent viscosities

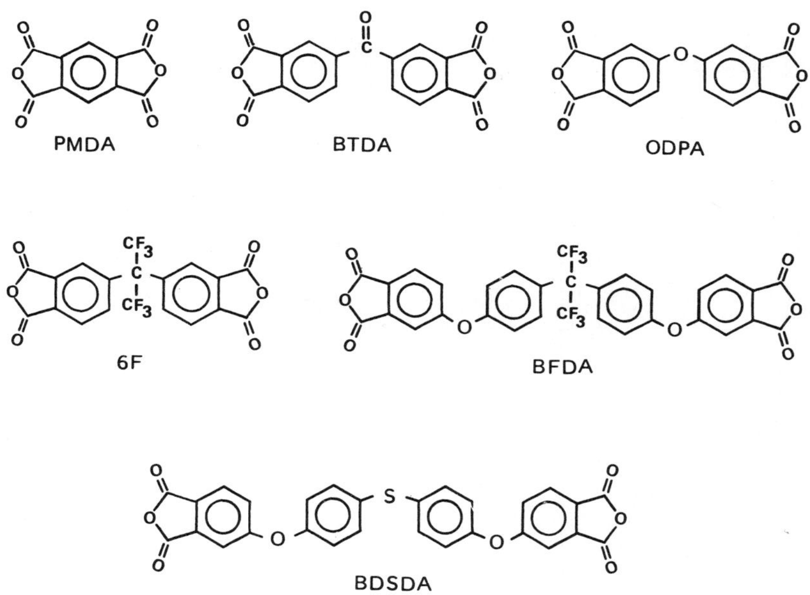

Fig. 1. - Aromatic dianhydrides used in the synthesis of polyimides.

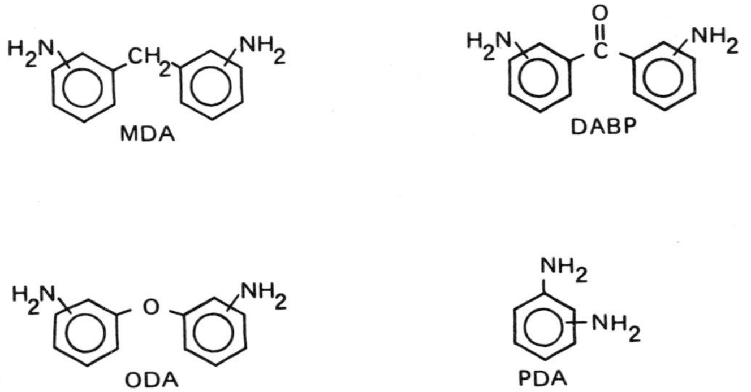

Fig. 2. - Aromatic diamines used in the synthesis of polyimides.

for each series of polyamic acids are listed in Tables I-VI. Films of the polyimides varied in color from yellow to nearly colorless after heating to 300°C (Tables I-VI). A thermal treatment of only 200°C was previously determined by Bell (10) to be insufficient for complete ring closure and/or drying for a series of all benzophenone polyimides, whereas films heated to 300°C appeared to be completely imidized. Infrared spectra of films cured to 300°C displayed characteristic imide related bands at 1770-1790, 1710-1735, and 742-747 cm^{-1}. All films were clear with the exception of the BDSDA-*p*-PDA film which was opaque and cream-colored in appearance, and exhibited a high degree of crystallinity (14).

The thermal conversion from polyamic acid to polyimide resulted in a decrease in inherent viscosity from 0.77 to 0.38 for ODPA-*o*,*p*'-ODA, 0.75 to 0.28 for 6F-*o*,*p*'-ODA and 1.09 to 0.54 for BDSDA-*o*,*p*'-ODA. However, these imide solutions, when cast and dried to 300°C, produced flexible films. Chemical conversion of ODPA-*o*,*p*'-ODA polyamic acid to polyimide at room temperature resulted in a similar decrease in the inherent viscosity (0.52 - 0.38). A similar viscosity relationship between polyamic acids and soluble polyimides was first noted by Angelo (15), and has more recently been reported by Harris and Norris (3) for phenylated polyimides. The higher viscosities of the polyamic acids might be due to *inter*molecular hydrogen bonding between the amide and carboxylic acid groups. Other factors such as *intra*molecular H-bonding or H-bonding of the amic acids with solvent molecules should be considered in the overall effect. In any case, the extent of hydrogen bonding of any kind should be

Table I. ODPA Polyimide Properties

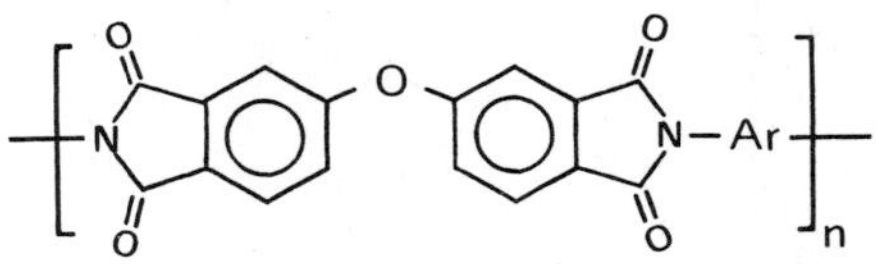

Ar diamine	η_{inh} [b]	T_g (°C) [c]	Film Solubility [a] NMP	DMF	m-cresol
m,m'-DABP	0.66	248	i	i	i
o,p'-DABP	0.11	265 (TBA)	s	s	sp
m,p'-DABP	0.38	263	i	i	i
p,p'-DABP	0.74	280	i	i	i
m,m'-MDA	0.78	258	i	i	i
o,p'-MDA	0.66	279 (263 N_2)[d]	i (s)	i (s)	i (s)
m,p'-MDA	0.92	274	i	i	i
p,p'-MDA	0.89	294 (275 N_2)[d]	i (s)	i (i)	i (s)
o,p'-ODA	0.77	264	s	s	s
p,p'-ODA	0.77	270	s	i	s
m-PDA	0.78	313	i	i	i
p-PDA	1.82	342	i	i	i

[a] s = soluble, sp = sparingly soluble, i = insoluble.

[b] Inherent viscosities of polyamic acids were run at 0.5% in DMAc at 35°C.

[c] T_gs were measured by thermomechanical analysis.

[d] Films cured in nitrogen.

Table II. 6F Polyimide Properties

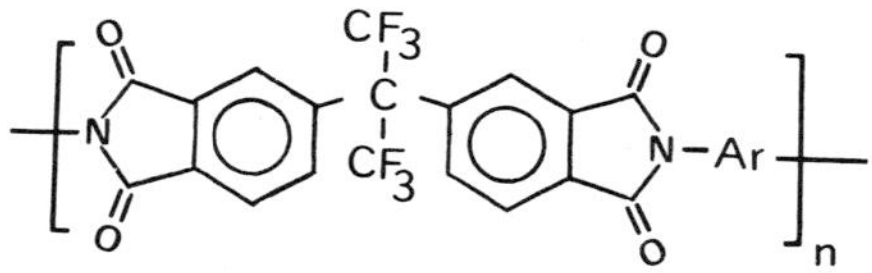

Ar diamine	η_{inh}[b]	T_g (°C)[c]	Film Solubility[a] NMP	DMAc	DMF	m-cresol
m,m'-DABP	0.47	260	s	s	s	sp
o,p'-DABP	0.10	289 (TBA)	s	s	s	s
m,p'-DABP	0.25	288	s	s	s	i
p,p'-DABP	0.35	311	i	i	i	i
m,m'-MDA	1.23	248 ($242\ N_2$)[d]	i (s)	i (s)	i (s)	i (sp)
o,p'-MDA	0.55	245 ($262\ N_2$)[d]	s (s)	sp (s)	sp (s)	s (s)
m,p'-MDA	0.86	272	i	i	i	i
p,p'-MDA	1.82	290	i	i	i	i
o,p'-ODA	0.75	276	s	s	s	s
p,p'-ODA	2.20	307	i	sp	s	s
m-PDA	0.94	303	s	s	s	s
p-PDA	0.90	339	s	s	s	s

[a] s = soluble, sp = sparingly soluble, i = insoluble.

[b] Inherent viscosities of polyamic acids were run at 0.5% in DMAc at 35°C.

[c] T_gs measured by thermomechanical analysis unless indicated otherwise.

[d] Films cured in nitrogen.

Table III. BFDA Polyimide Properties[a]

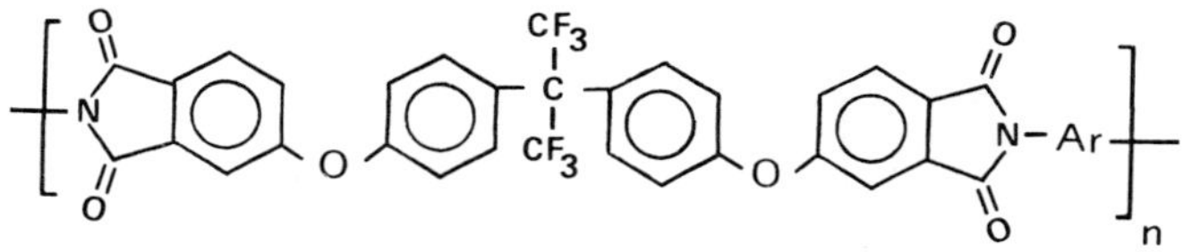

Ar diamine	η_{inh}[b]	T_g (°C)[c]
m,m'-DABP	0.57	220
o,p'-DABP	0.13	234
m,p'-DABP	0.30	237
p,p'-DABP	0.41	248
m,m'-MDA	0.63	216
o,p'-MDA	0.33	224
m,p'-MDA	0.48	231
p,p'-MDA	0.45	238
o,p'-ODA	0.35	235
p,p'-ODA	0.58	237
m-PDA	0.44	240
p-PDA	0.98	254

[a] All polymers were insoluble in organic solvents.

[b] Inherent viscosities of polyamic acids were run at 0.5% in DMAc at 35°C.

[c] T_gs measured by thermomechanical analysis.

Table IV. BDSDA Polyimide Properties

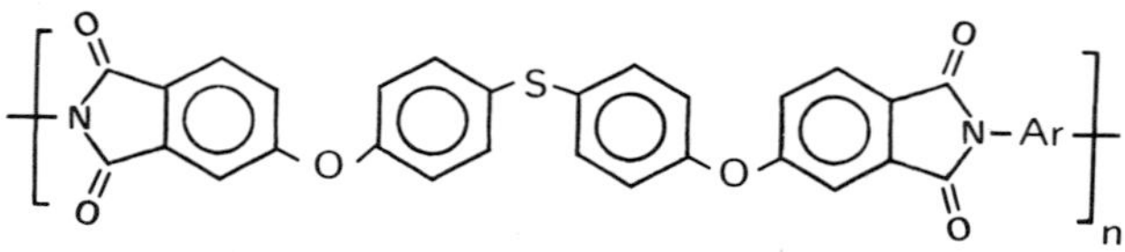

Ar diamine	η_{inh} [b]	T_g (°C) [c]	Film Solubility[a] NMP	DMAc	DMF	m-cresol
m,m'-DABP	1.09	200	s	i	i	sp
o,p'-DABP	0.54	212	s	s	s	s
m,p'-DABP	0.88	213	s	i	i	sp
p,p'-DABP	1.25	221	s	i	i	s
m,m'-MDA	1.12	247	i	i	i	i
o,p-MDA	0.68	240 ($223\ N_2$)[d]	i (s)	i (sp)	i (s)	i (s)
m,p'-MDA	1.27	254	i	i	i	i
p,p'-MDA	1.19	267 ($223\ N_2$)[d]	i (sp)	i (sp)	i (i)	i (i)
o,p'-ODA	1.09	216	s	sp	s	s
p,p'-PDA	1.30	221	s	i	i	s
m-PDA	1.28	224	s	i	i	i
p-PDA	1.54	226(422)[e]	i	i	i	i

[a] s = soluble, sp = sparingly soluble, i = insoluble.

[b] Inherent viscosities of polyamic acids were run at 0.5% in DMAc at 35°C.

[c] T_gs were measured by thermomechanical analysis.

[d] Films cured in nitrogen.

[e] DSC on this opaque film showed an intense endotherm at 422°C indicating a melt.

Table V. BTDA Polyimide Properties

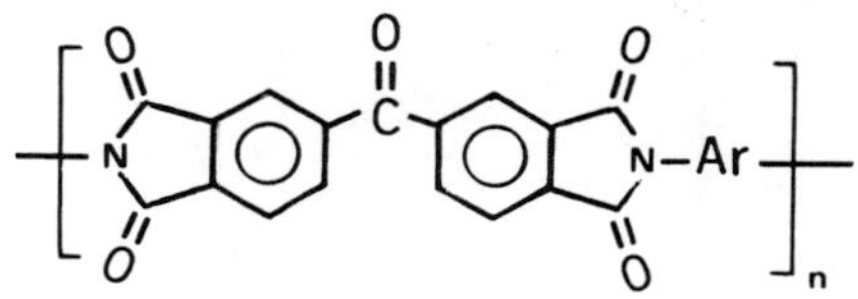

Ar diamine	η_{inh}[b]	T_g (°C)[c]	Film Solubility[a] NMP	DMF	m-cresol
m,m'-DABP	0.70	259	i	i	sp
o,p'-DABP	<0.10	—	—	—	—
m,p'-DABP	0.64	277	i	i	i
p,p'-DABP	0.79	288	i	i	i
m,m'-MDA	0.73	265	i	i	i
o,p'-MDA	0.62	290	sp	sp	sp
m,p'-MDA	0.99	281	i	i	i
p,p'-MDA	0.83	296	i	i	i
o,p'-ODA	0.60	281	sp	sp	sp
p,p'-ODA	1.80	280	i	i	i
m-PDA	0.86	300	i	i	i
p-PDA	1.31	333	i	i	i

[a] s = soluble, sp = sparingly soluble, i = insoluble.

[b] Inherent viscosities of polyamic acids at 0.5% in DMAc at 35°C.

[c] T_gs were measured by thermomechanical analysis.

Table VI. PMDA Polyimide Properties

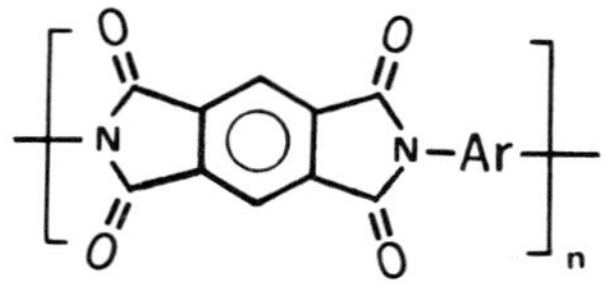

Ar diamine	η_{inh}[b]	T_g (°C)[c]	Film Solubility[a] NMP	DMF	m-cresol
m,m'-DABP	0.83	326	i	i	i
o,p'-DABP	0.17	360 (TBA)	s	s	sp
m,p'-DABP	0.84	347	i	i	i
p,p'-DABP	0.69	412	i	i	i
m,m'-MDA	0.99	308	i	i	i
o,p'-MDA	0.60	347 (TBA)	s	sp	s
m,p'-MDA	0.89	314	i	i	i
p,p'-MDA	1.51	339	i	i	i
o,p'-ODA	0.52	366 (TBA)	s	s	s
p,p'-ODA	1.46	399	i	i	i
m-PDA	1.55	442	i	i	i
p-PDA	1.49	—	i	i	i

[a] s = soluble, sp = sparingly soluble, i = insoluble.

[b] Inherent viscosities of polyamic acids at 0.5% in DMAc at 35°C.

[c] T_gs were measured by thermomechanical analysis unless indicated otherwise.

greatly reduced when the amic acid system is imidized, thus reducing the viscosity of the polyimide.

B. Thermomechanical Properties

Glass transition temperatures (T_g) of the polyimide films preheated to 300°C were measured by thermomechanical analysis (TMA). The T_g data for each series of polymers is listed in Tables I-VI. Good correlation in T_g values of films cured in air and in nitrogen at 300°C was observed except for those polymers made from diphenylmethane diamines. In most instances the MDA films heated in air were less flexible and had higher T_gs than those heated in nitrogen, which would indicate that oxidative crosslinking had occurred. Air-induced crosslinking in polyimides containing a methylene group had been noted previously (10, 16, 17). In several cases where polymer films were too brittle for measurement by TMA, thermomechanical transitions were determined by torsional braid analysis (TBA). The extreme fragility of the ODPA, 6F and PMDA polymers containing *o,p'*-DABP was most probably due to low molecular weights. The inherent viscosity of BDSDA-*o,p'*-DABP was unusually high compared to other polymers containing this particular diamine.

Lower T_gs were observed for *meta*-substituted polymers in comparison to the *para*-substituted polymers studied. Similar results relatable to the isomerism of the aromatic diamine have been described for other polyimide systems (10, 18). For polymers in Tables I-VI, inclusion of *o,p'*-diamines into the polymer backbone produced polymers with lower T_gs than those containing all-*para* diamines, but not as low as for the all-*meta* polymers. Similar effects on the melting behavior of addition polyimide oligomers containing *meta* and *o,p'*-diamines of MDA and DABP have been noted (19). The *o,p'*-diamines, however, were not as effective in lowering the T_gs of some BTDA and PMDA polymers (10). In fact, BTDA-*o,p'*-MDA had T_gs as high or higher than the all-*para* polymers (Tables V and VI).

C. Polymer Solubility

The solubilities of polyimide films cured at 300°C are presented in Tables I-VI. Films were tested for solubility at 2% solids (w/w) in NMP, DMF, *m*-cresol and in some instances DMAc. ODPA, 6F and BDSDA anhydrides produced polyimides with greater solubilities than for similar polyimides containing the dianhydrides BTDA and PMDA (2, 10). Air cured polymers of ODPA, 6F and BDSDA containing MDA diamines were generally insoluble, but the same polymers proved to be soluble when imidized in nitrogen, thus minimizing air-induced crosslinking. All BFDA polyimides (Table III) were unexpectedly insoluble in the above solvents with evidence of swelling. Even films of the BFDA polymers containing *o,p*'-ODA were totally insoluble, whereas all the other polyimides made from this particular diamine were soluble. Additional solubility tests were performed on BFDA polyimide films in fuming nitric acid. After a 1 min. time period in HNO_3, only BFDA-*o,p*'-DABP was sparingly soluble and all other BFDA films were insoluble. After 15 minutes, BFDA-*o,p*'-DABP was soluble, BFDA-*o,p*'-MDA was sparingly soluble and all others remained insoluble. Therefore, the BFDA polymers were assumed to be crosslinked.

A second technique which led to soluble polyimides was that of varying the positions at which amino groups were attached to the aromatic rings relative to the bridging groups connecting those aromatic rings. Those polymers containing *o,p*'-diamines were almost always soluble. *Ortho* substitution consistently gave more soluble polymers than did *meta*-substitution even though the T_gs of *o,p*'-oriented polymers were higher. The solubility of polymers made from PMDA and *o,p*'-diamines was most surprising, in that heretofore PMDA polyimides have been considered to be insoluble (2). This unique enhanced solubility of the *ortho*-related polymers might be attributed to the higher entropy state of the polymer chains resulting from the *ortho* linkages. The greater disorder should facilitate solvation of the polymer chains.

Films of ODPA-*o,p*'-ODA, 6F-*o,p*'-ODA, and BDSDA-*o,p*'-ODA were further tested at 15% solids (w/w) in DMF and found to be completely soluble at room temperature. These 15% polyimide solutions, which were cast onto glass plates and heated 1 hour at 300°C, yielded flexible films with the same as or slightly higher T_gs than the original films (Table VII). Additional testing of ODPA-*o,p*'-ODA, which had been chemically (rather than thermally) converted from the polyamic acid to polyimide, proved this polymer to be readily soluble to at least 40% solids (w/w) in DMF at room temperature. This 40% polyimide solution, which was cast onto plate glass and cured at 300°C for 1 hour, produced a clear yellow, flexible film with a T_g of 262°C compared to 264°C (Table I) for the ODPA-*o,p*'-ODA film cast from the polyamic acid and thermally imidized. The T_g of this soluble polymer remained essentially unchanged regardless of the method used for imidization.

Table VII. T_g Comparison Of Cured Films Cast From Polyamic Acid Solutions VS Polyimide Solutions

Polymer	Polyimide η_{inh}	Film from polyamic acid soln. cured to 300°C T_g	Film from 15% imide soln., dried at 300°C T_g
ODPA - o,p'-ODA	0.38	264	265
6F - o,p'-ODA	0.28	276	285
BDSDA - o,p'-ODA	0.54	216	223

D. Thermal Properties

Dynamic TGAs of polyimide films containing *p,p'*-DABP as the aromatic diamine moiety and presented in Figure 3 and

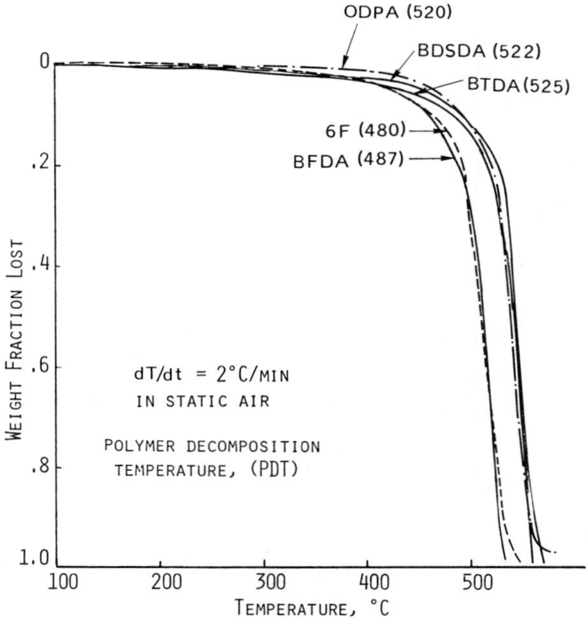

Fig. 3. - Thermograms of p,p'-DABP polyimide films cured to 300°C in air.

compared with that of BTDA-*p*,*p*'-DABP (10). Polymer decomposition temperatures (PDT) shown in parenthesis were obtained as the intersection of tangents drawn to the zero and maximum weight loss portions of the TGA curves. From these data, ODPA and BDSDA polymers were as thermally stable as the polyimide containing BTDA and, surprisingly, somewhat more stable than those made from the perfluorinated dianhydrides, 6F and BFDA. Isothermal TGAs at 316°C on polyimides made with *p*-PDA are shown in Figure 4. After 500 hours at temperature, the 6F polymer exhibited the greatest thermooxidative stability, which confirms the long-accepted recognition of 6F-derived polyimides as having outstanding isothermal oxidative stability.

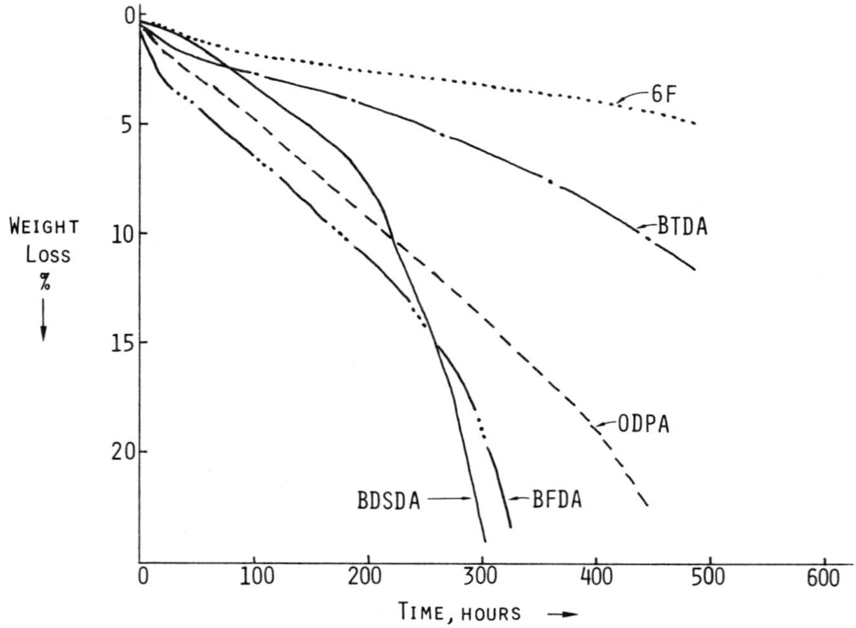

Fig. 4. - Isothermal weight loss at 316°C in air of p-PDA polyimide films cured to 300°C.

IV. CONCLUSIONS

A structure-solubility study was conducted on a series of high molecular weight, condensation polyimides prepared by reacting four aromatic dianhydrides with isomeric aromatic diamines. The solubilities of these polyimides were compared with those of polymers made from BTDA and PMDA. Both modification of bridging groups in the dianhydride structures and variation of the isomer position of the diamine monomers effected polyimide solubility. In addition, the reaction of *ortho* and *meta*-oriented diamines with ODPA, 6F, BDSDA and BFDA polymers resulted in polyimides with lower glass transition temperatures than those from the rigid *para* diamines. Exploitation of these methods for making soluble, high molecular weight polyimides can be expected to increase the potential value of the polyimides by improving their processability.

V. REFERENCES

1. Adrova, N.A., Bessonov, M.I., Laius, L.A. and Rudakov, A.P., "Polyimides, A New Class of Thermally Stable Polymers," Technomic Publishing Co., Stamford, Conn., p. 89, 1970.
2. Sroog, C.E, Endrey, A.L., Abramo, S.V., Berr, C.E., Edwards, W.M., and Olivier, K.L., J. Polym. Sci., A, 3, 1373 (1965).
3. Harris, F.W. and Norris, S.O., J. Polym. Sci., A-1, 11, 2143 (1973).
4. Korshak, V.V., Vinogradova, S.V. and Vygodskii, Y.S., J. Macromol. Sci. - Rev. Macromol. Chem., C-11 (1), 45 (1974).
5. Harris, F.W., Feld, W.A. and Lanier, L.H., Polymer Letters Edition, 13, 283 (1975).
6. Kolesnikov, G., Fedotova, O., Hofbauer, E. and Shelgayeva, V., Vysokomol. Soedin., A9, 612 (1967).
7. Coe, D.G., U.S. Patent 3,310,573 (to E.I. duPont de Nemours and Co.), March 21, 1967.
8. Vaughan, R.W., Jones, R.J., O'Rell, M.K. and Zakrezewski, G.A., TRW Systems Group, Redondo Beach, CA, 90278, Contract No. NAS3-17824.
9. Takekoshi, T., Kochanowski, J.E., Manello, J.S. and Webber, M.J., General Electric Corporate Research and Development, Schenectady, NY. 12301, Contract No. NAS1-12079.
10. Bell, V.L., Stump, B.L. and Gager, H., J. Polym. Sci., A-1, 14, 0000 (1976).
11. Burks, H.D., J. Appl. Polym. Sci., 18, 627 (1974).
12. Dalal, S.K., Carl, G.L., Inge, A.T. and Johnston, N.J., paper presented at the 167th Meeting, A.C.S., Los Angeles, CA., April 1974; Polymer Preprints, 15 (1), 576 (1974).

13. Jewell, R.A., J. Appl. Polym. Sci., 15, 1717 (1971).
14. St.Clair, T.L. and St.Clair, A.K., J. Polym. Sci., Polym. Lett. Ed., in press.
15. Angelo, R.J., U.S. Patent 3,546,175 (to E.I. duPont de Nemours and Co.), December 8, 1970.
16. Jewell, R.A. and Sykes, G.F., paper presented at the 173rd Meeting, A.C.S., San Francisco, CA., September 1976; Div. Coatings and Plastics Chem., Preprints, 36(2), 258 (1976).
17. Bell, V.L., J. Polym. Sci., A-1, 14, 225 (1976).
18. Bell, V.L., paper presented at the 165th Meeting, A.C.S., Dallas, TX April 1973; Org. Coatings and Plastics Chem., Preprints, 33, 153(1973).
19. St.Clair, A.K. and St.Clair, T.L., Polymer Engineering and Science, 16(5), 314(1976).

SOLUBILITY PROPERTIES OF ORGANOMETALLIC CONDENSATION POLYMERS

Charles E. Carraher, Jr.
Wright State University

I. INTRODUCTION

This paper contains a brief review of the solubility properties of organometallic condensation polymers and relates possible explanations for the observed results.

II. GENERAL SOLUBILITY PROPERTIES

Relative to more classical polymers, organometallic polymers of form I are soluble in

$$\text{-A-M-A-R-}$$
$$\text{I}$$

fewer solvents and to lesser extents within these solvents (for instance 1,2). In fact many are insoluble in all solvents attempted. These poor solubility properties are probably due to a combination of factors including a) the crystalline nature of many of such products, b) high cohesive (secondary bonding) forces between chains, and c) a peculiar combination of bonding offered by the polymers - a combination of both nonpolar and high to low polar contributions. These factors are related.

Briefly considering overall polymer solubility, while the entropy term favors solubility, it is normally opposed by the enthalpy term. According to Hildebrand, Scott, etc., one attempts to generate small enthalpy of mixing values to effect polymer solution. Since many organometallic polymers contain a combination of polar (ranging from high to low) and nonpolar portions a good solvent might also be expected to exhibit both polar and nonpolar regions. Solvents which fit this criteria include the dipolar aprotic solvents as DMSO[1], DMF[2], HMPA[3] and acetone. In fact these solvents are generally the best solvents for organometallic polymers.

Solvents which have successfully solubilized several polymer families are DMSO, DMF, HMPA, TEP[4] and acetone. A number of other solvent systems have been suggested but have

1. DMSO = dimethylsulfoxide
2. DMF = N,N-dimethylformamide
3. HMPA = hexamethylphosphoamide
4. TEP = triethylphosphate

yet to successfully dissolve a number of polymers. These include organic liquid-salt combinations such as NaI-acetone, LiCl-acetone and LiCl-DMF; HCA; and esters and amides as diethylmalonate.

Another ploy utilized to attempt solubilization of a number of organometallic polymers is heating the polymers to above their Tg in the test solvent systems. While this has proven successful in select systems, many of the products undergo initial degradation via solid-gas transitions before the Tg is reached. Thus caution must be exercised in attempting solubilization of the polymers utilizing heating. For instance, products from the condensation of uranyl nitrate hexahydrate with aromatic diacids are insoluble in all solvents at room temperature but at 70°C become soluble in dipolar aprotic solvents as DMSO. Solubilization is accompanied by loss of two waters (3).

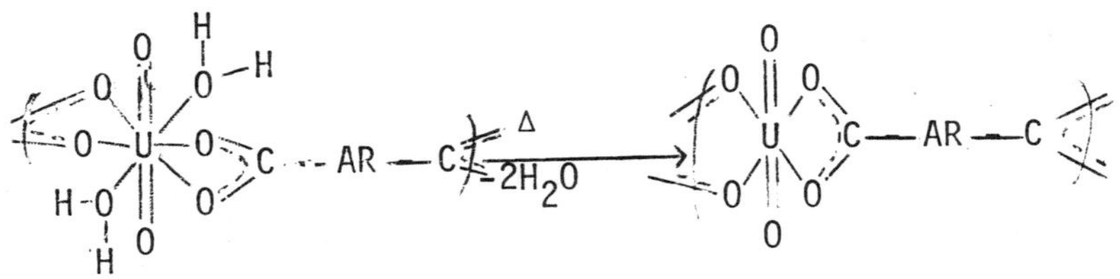

We have also employed nucleophilic solvents as 2-chloroethanol and 2,2,2-trifluoroethanol (for instance 1,2,4-8). In every case where solubilization appeared to occur, degradation occurred. In select cases where solubility was effected rapidly (within 6 hours or less) M_w's were obtained within the 10^5 to 10^7 range being reduced to ca 10^3 after ca several hours to several days (c.f. Table 1). It is strongly suspected that

TABLE 1
Degradation of poly[bis(oxycarbonylcyclopentadienyl)cobalt-(III)(dicyclopentadienyltitan(IV)] hexafluorophosphate as a function of time.

Time[a.] (mins)	$\eta_{sp/c}$[b.] (ml/g)	$\bar{M}_w$[c.] (10^{-4})
25	22	8.1
60	15	2.1
90	—	1.6
180	8	1.1

a. Time from initial mixing of product with 2-chloroethanol. Solubilization is rapid and effected in 15 mins. with rapid vibration.

b. Concentration is 0.1%.

c. From light scattering measurements utilizing the dissymmetry method for single measurements on 0.1% solutions.

degradation accompanies the solubilization of a number of organometallic polymers, but this has yet to be unambiguously tested since those particular products are insoluble in most other solvents.

A number of acid solutions have been reported as being successfully employed to effect solubilization of metal containing polymers. We find that with few exceptions degradation results rather than the desired polymer solubilization in such acid solutions - namely formic, acetic, sulfuric and nitric acids.

For the three elemental families studied range and extent of solubility decreases with increase in the size of the metal (solubility Si>Ge>Sn>Pb; (tentatively) Ti>Zr>Hf; and (tentatively) As>Sb>Bi; for instance 1, 2, 9, 10). The reason(s) for this trend is not known but may involve increased - longer distance secondary bonding by the larger atoms between chains thus increasing the cohesive forces between chains.

One area of future study related to achieving better solubility is the disruption of polymer symmetry, which may decrease polymer crystallinity by use of unsymmetrical, large substituants on either or both of the co-monomers.

As an additional point of interest, the high amount of polar bonding present within many of the organometallic polymers may result in the interesting and potentially useful phenomenon referred to as "Anomalous Fiber Formation" which occurs with a number of titanium containing products, but not the corresponding zirconium or hafnium products, presumably because of the greater impact of interchain bonding effected by the smaller titanium moiety (though this must be viewed with caution because of the nearness in size between titanium and zirconium) (for instance 5-8, 10, 33-39). Solubility between fiberous and nonfiberous materials from the same reaction differ. For instance nonfiberous titanium polyesters go to a gel state in many dipolar aprotic solvents whereas the corresponding fiberous materials remain unaffected (34).

III. UNUSUAL VISCOSITY:LIGHT SCATTERING-MOLECULAR WEIGHT RATIOS

One objective for the synthesis of new polymers is the evaluation of their obedience to general polymer theories. The relatively large drag effected in dilute solution resulting from the polymer chains residing in several flow planes is one characteristic of polymers often cited.

A number of organometallic polymers synthesized by us exhibit unusually low viscous drays compared to their molecular weight determined by light scattering photometry (cf. Table 2). This may be due to at least three factors. The first factor concerns anomalous scattering due to color or absorption. Utilizing the prescribed procedures we have not found where this is a factor for our products. The last two factors can both be related to the "stiffness" of many of the organometallic polymers.

TABLE 2
LVN: M_w, K and a values for selected polymers[a].

Polymer (Name or Structure)	Solvent	LVN/ $\bar{M}_w(10^4)$	$K(10^3)$	a	Ref.
Polystyrene	Benzene	8.4	9.2	0.74	12
Polyethylene	p-Xylene	333	16	0.83	13
Polypropylene	Decalin	17	11	0.80	14
Polyacrylonitrile	DMF	24	24	0.75	15
Poly(vinylchloride)	Cyclohexanone	27	170	0.55	16
Poly(ethylene terephthalate)	Phenol/Tetrachlorophenol	25	47	0.68	17
Poly(hexamethylene adipamide)	90% aqueous formic acid-2M KCl	24	140	0.56	18
Poly(dimethyl siloxane)	Bromobenzene	7.6	76	0.5	19
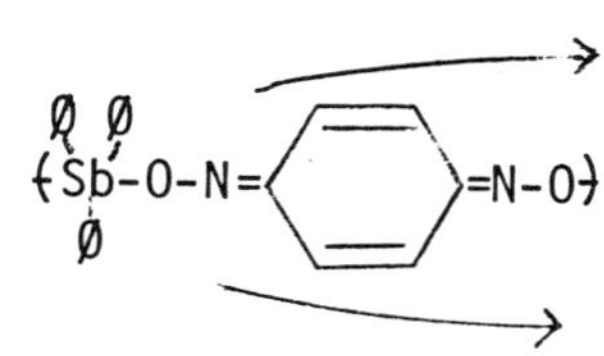	Hexamethylphosphoramide	0.087	2.2	0.4	11
	Triethyl phosphate	0.14			11

(Table 2 Cont'd.)

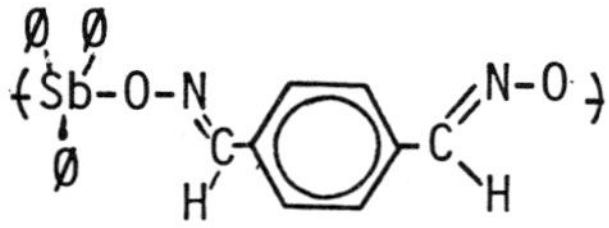	DMSO	0.071	.71	.5	11
	Sulfolane-W	0.70			11
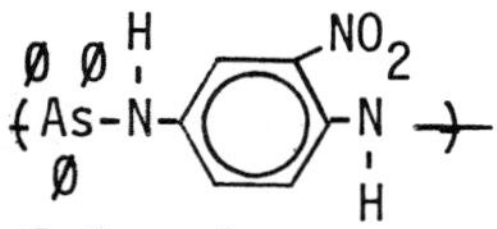	DMSO	0.40			20
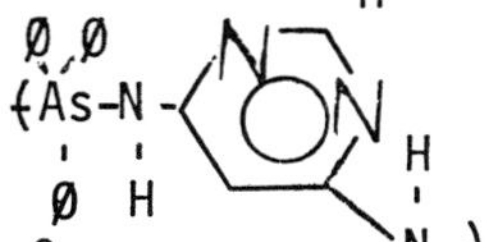	DMSO	0.14			20
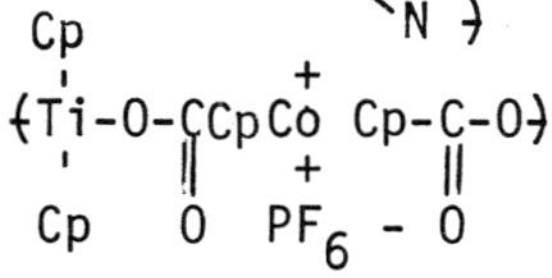	2-Chloro-ethanol	5.0	15	0.63	8
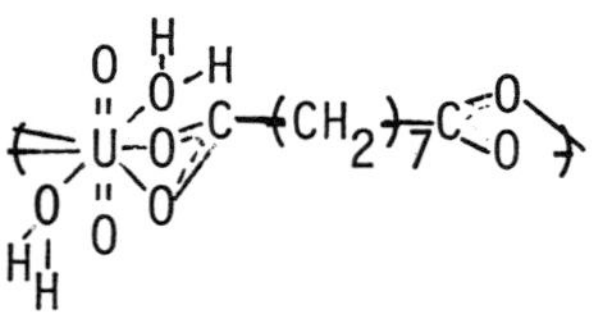	DMSO	48	300	0.55	3
Poly(sodium acrylate)	1.5M Aqueous NaBr	12	124	0.5	23
Poly(benzyl methacrylate)	Benzene	2.0	1.03	0.82	24
Poly(4,4'-isopropylidenecarbonate)	Chloroform	28	277	0.5	25
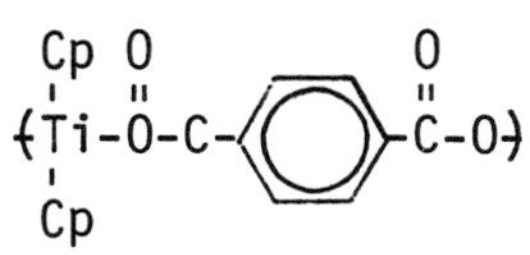	2-Chloro-ethanol	1.8	21	0.48	22
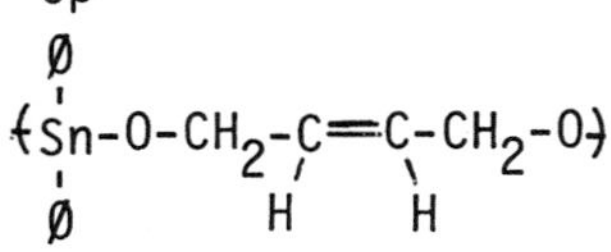	1-Octanethiol	0.25	6.2	0.4	21

(Table 2 Cont'd.)

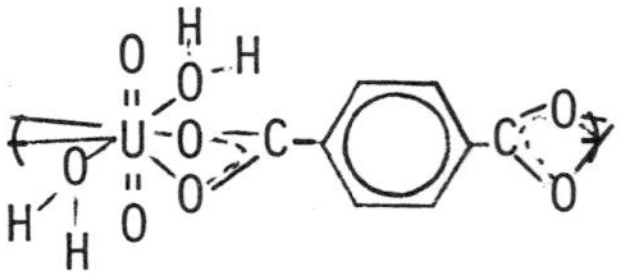	triethyl phosphate	0.73	25	0.53	3
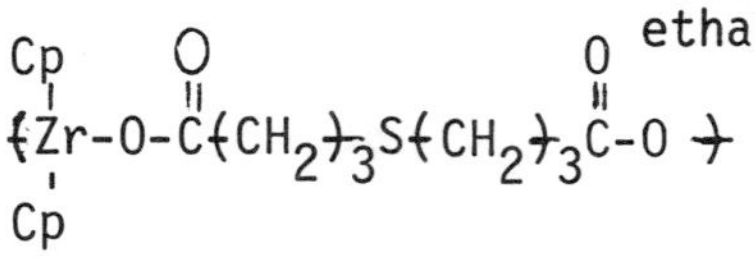	2-Chloro-ethanol	1.2			10
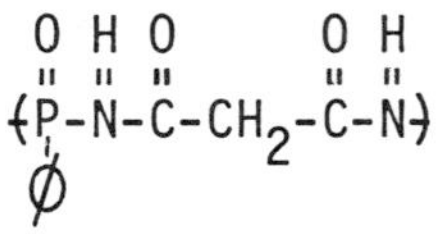	DMSO	0.22			22
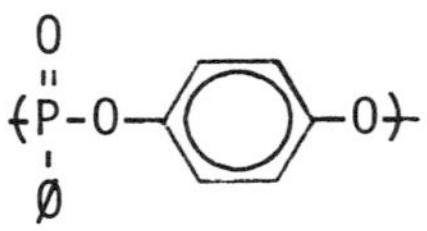	DMSO	5.5	20	0.61	10

a. $\bar{M}_w$ determined by light scattering photometry. The ratio LVN/$\bar{M}_w$ is "normalized" using $\bar{M}_w = 10^4$. LVN and K are in units of ml/g.

Many exhibit resonance throughout (large segments) the chain leading to what is called fluorescence whereby light is absorbed and emitted equally in all directions unlike "pure" scattered light where scattering envelopes are formed with such chain resonance must be corrected for fluorescence.

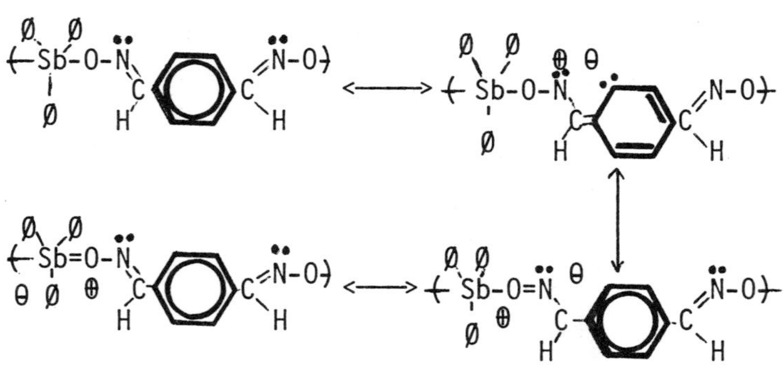

For instance the product from triphenylantimony dichloride with terephthaldicarboxaldehyde dioxime (pictured above) exhibited a $\bar{M}_w$ value of 1.5×10^8 before correction for fluorescence, with $\bar{M}_w$ reduced to 6.9×10^5 after correction (11). All results listed in Table 2 are corrected for fluorescence. Even after correction for fluorescence the products exhibited unusually low LVN/M_w ratios. Also products which did not exhibit fluorescence but do possess restricted motion exhibit the unusually low ratios. Both resonance and sterically restricted motion contribute to an overall chain stiffness. This semirigid nature may permit the chains to be more easily accommodated within a small number of sheer planes resulting in a relatively low viscous drag.

Chain stiffness alone is not responsible for the low drag observed (cf. Tables 2, 3). Neither is polymer insolubility alone responsible for the low drag and unusually low LVN/$\bar{M}_w$ ratios. The latter is based on data such as given in Table 3 where the LVN/$\bar{M}_w$ ratio is approximately constant for a number of polymer systems when comparing theda with nontheda polymer-solvent combinations. It presently appears that it takes both chain stiffness and poor solubility to effect the observed low drags with both factors acting to force the polymer chains into a small number of flow planes - possibly between planes.

The factor of poor solubility is emphasized in the following data. A product from the condensation of triphenylantimony (V) dichloride with terephthaldicarboxaldehyde dioxime near its theda point in DMSO exhibits very little slope in plots of η sp/c (k = 0.006) and ln η rel/c (k = 0.004) verses concentration (calculated utilizing concentrations in g/dl) yielding a LVN of 4 ml/g. In sulfolane-W, a better solvent as indicated by greater slopes in the η sp/c (k = 0.22) and ln η rel/c (k = 0.1), the LVN increased to 39 ml/g.

TABLE 3
LVN/M_w Ratios For Theda and Nontheda Systems[a].

Polymer	$K \cdot 10^3$	a	LVN/$\bar{M}_w$	Ref.
Poly(1-butene) θ	123	0.50	12	26
Non θ	7.3	0.80	12	26
Poly(styrene) θ	80	0.50	8	27
Non θ	22	0.64	8	24
Poly(vinyl chloride) θ	156	0.50	16	28
Non θ	16	0.77	19	29

(Table 3 Cont'd.)

Poly(vinyl acetate) θ	82	0.50	8	24
Non θ	32	0.65	12	30
Poly(methyl acrylate) θ	54	0.50	5.4	23
Non θ	4.5	0.78	5.9	24
Poly(tetrahydrofuran) θ	206	0.49	19	24
Non θ	42	0.65	17	24
Nylon 66 θ	253	0.50	25	31
Non θ	142	0.56	25	31

a. Ibid a in Table 2.

Table 2 also contains a listing of Mark-Houwink "K" and "a" values. Neither the "a" or "K" values alone appear to account for the unusual LVN/M_w ratio but considered taken as a pair lower for most of the organometallic polymers when compared with typical organic polymers. The low "a" values probably results from the characteristic poor solubility of the organometallic polymers. Further subdivision of the K terms in accordance with the Flory-Fox approach leads to the following:

$$K = \phi \frac{\bar{r}_o^{\,2}}{M}$$

It is possible that the low K values results from the fact that for a particular $\bar{r}_o$ unit there exists a greater M (relative to more typical organic polymers) due to the presence of the heavier metal atoms. This has yet to be tested but should be demonstratable by synthesis of analogous families of organometallic polymers where only the M is varied and then comparing their K values. Because of solubility problems and the seemingly inability to synthesize select organometallic polymers with wide variances in molecular weight we have yet to investigate this.

While the presence of the inorganic moiety might also have an effect on the solution behavior of the organometallic polymers beyond that contributing towards poor solubility and chain stiffness, it is currently considered to be of a secondary nature. For instance the product from the condensation of uranyl nitrate hexahydrate with disodium azelate can be considered flexible, and is fairly soluble in a number of dipolar aprotic solvents exhibiting LVN/$\bar{M}_w$ ratios typical of organic polymers (3). Yet the analogous product from disodium terephthalate (which should be "stiffer") exhibits a low LVN/M_W ratio characteristic of many of the stiffer organometallic polymers.

Stiffness may also act to decrease the rate of solubilization since chain flexibility can permit inclusion of solvent molecules within the inner portions of the dissolving polymer.

IV. REFERENCES

1. Carraher, C., "Interfacial Synthesis," Dekker, 1976, Chapter 11 (E. Millich and C. Carraher, Eds.).
2. Carraher, C., Inorganic Macromolecules Reviews, 1, 271 (1972).
3. Carraher, C., and Schroeder, J., Polymer Letters 13, 215 (1975) and Polymer Preprints, 16, 659 (1975) and unpublished results.
4. Carraher, C., and Dammeier, R., Makromolekulare Chemie, 135, 107 (1970).
5. Carraher, C., J. Polymer Sci., A-1, 9, 3661 (1971).
6. Carraher, C., and Reimer, J., Polymer (Br.), 13, 153 (1972) and J. Polymer Sci., A-1, 10, 3367 (1972).
7. Carraher, C., and Bajah, S., Polymer (Br.), 14, 42 (1973); 15, 9 (1974) and British Polymer J., 7, 321 (1975).
8. Carraher, C., and Sheats, J., Makromolekulare Chemie, 166, 23 (1973).
9. Carraher, C., and Dammeier, R., J. Polymer Sci., 10, 413 (1972).
10. Carraher, C., Angew. Makromolekulare Chemie, 28, 145 (1973); Organic Coatings and Plastics Chemistry, 33, 629 (1973) and Makromolekulare Chemie, 166, 31 (1973) and unpublished results.
11. Carraher, C., and Hedlund, L., paper in progress.
12. Orofino, T., and Wenger, F., J. Phys. Chem., 67, 566 (1963).
13. Trementozzi, Q., J. Polymer Sci., 36, 113 (1959).
14. Kinsinger, J., and Huges, R., J. Phys. Chem., 63, 2002 (1959).
15. Cleland, R., and Stockmayer, W., J. Polymer Sci., 17, 473 (1955).
16. Moore, W., and Hutchinson, R., Nature, 200, 1097 (1963).
17. Moore, L., Polymer Preprints, 1, 234 (1960).
18. Elias, H. G., and Schumacher, R., Makromol. Chemie, 76, 23 (1964).
19. Schulz, G., and Haug, A., Z. Physik. Chem. (Frankfurt), N. F., 34, 328 (1962).
20. Carraher, C., and Moon, W., unpublished results.

21. Carraher, C., and Scherubel, G., Makromol. Chemie, 160, 259 (1972).
22. Carraher, C., Org. Coat. and Plastics Chemistry, 31, 330 (1971).
23. Carraher, C., and Winthers, D., J. Polymer Sci., A-1, 2417 (1969).
24. Takahashi, A., Kamei, T., and Kagawa, I., Mippon Kagaku Zasshi, 83, 14 (1962).
25. J. Brandrup and Immergut (Eds.), "Polymer Handbook," Interscience Pubs., N. Y., 1966, Section IV.
26. Chirico, A., Chim. Ind., 42, 248 (1960).
27. Krigbaum, W., Kurz, J., and Smith, P., J. Phys. Chem., 65, 1984 (1961).
28. Homma, T., Kawahara, K., Jujita, H., and Ueda, M., Makromolekulare Chemie, 67, 132 (1963).
29. Sata, M., Koshiishi, Y., and Asahina, M., J. Polymer Sci., B-1, 233 (1963).
30. Freeman, M., and Manning, P. P., J. Polymer Sci., A-2, 2017 (1964).
31. Kalpagam, V., and Rao, R., J. Polymer Sci., A-1, 223 (1963).
32. Elias, H., and Schumacker, R., Makromolekulare Chemie, 76, 23 (1964).
33. Carraher, C., and Nordin, R., J. Polymer Sci, A-1, 10, 521 (1972); Makromolekular Chemie, 164, 87 (1973).
34. Carraher, C., Chem. Tech., 741 (1972).
35. Carraher, C., and Lessek, P., Eup. Polymer J., 8, 1339 (1972).
36. Carraher, C., Eup. Polymer J., 8, 215 (1972).
37. Carraher, C., and Jambaya, L., Angew. Makromolekular Chemie, 39, 69 (1974) and 52, 111 (1976).
38. Carraher, C., and Nordin, R., J. Applied Polymer Sci., 18, 53 (1974).
39. Carraher, C., and Frary, R., J. Polymer Sci., A-1, 12, 799 (1974); Br. Polymer J., 6, 255 (1974).

PARTIAL SOLUBILITY PARAMETERS OF SOME POLYALKYLMETHYLSILOXANES

Ronald H. Baney, Carl E. Voigt and James W. Mentele
DOW CORNING CORPORATION

I. INTRODUCTION

Many of the unique properties of polydimethylsiloxane, such as its surface properties, are frequently attributed to its low cohesive energy density.(1) However, the influence of pendant side groups on siloxane chains on solubility and other properties attributed to cohesive energy density has not been well defined. This study was undertaken to determine this influence. A series of crosslinked polymers with the general formula $(RMeSiO)_n$ where R=Me, Et, or $CF_3CH_2CH_2$, were prepared and their swelling characteristics determined.

II. BACKGROUND

Siloxane polymers, $(RMeSiO)_n$, have bonds with more polarity than organic polymers (Table 1). The increased atomic size of silicon with respect to carbon and perhaps its ability to form d_π-p_π bonds to oxygen give rise to very open, flexible structures. Tables 2 and 3 list published bond angles and

TABLE 1

Bond lengths and polar character of $\left(\begin{matrix} CH_3 \\ | \\ Si\text{-}O \\ | \\ R \end{matrix}\right)_n$ *R=CH_3, CH_3CH_2, $CF_3CH_2CH_2$*

Element X	*Bond Lenghts (2) in Si-X*	*in C-X*	*Calculated % Ionic Character(3) in Si-X*	*in C-X*
Si	2.34	1.88Å	--	12
C	1.88	1.54	12	--
H	1.47	1.07	2	4
O	1.63	1.42	50	22

TABLE 2

Bond angles(1)

Si-O-Si	130°	C-C-C	112°
C-O-C	111°	H-C-H	109°
C-Si-C	109°		

TABLE 3

Barrier of rotation(4)

Z	*C-Z*	*Si-Z*
$-CH_3$	3.6kcal	1.6kcal
-O-	2.7	<0.2

barriers of rotation for bonds found in carbon backbone and siloxane structures. This open, flexible structure of silicones might be expected to give low cohesive energy density.

Various physical properties of this series of siloxane polymers are presented in Table 4. The energy of activation

TABLE 4

Physical properties of $(RMeSiO)_x$

R	CH_3-	CH_3CH_2-	$CF_3CH_2CH_2$-
E_{vis}	3.4kcal	4.5kcal	7.9*
T_g	-123°C	-135°C	-70°C*
Surface Tension	21.5 dynes cm^{-1}	23.1 dynes cm^{-1}	24.0 dynes cm^{-1}
Polar fraction to S.T.	0.41	0.39	0.52
Thermal Expansion $\times 10^4 -\frac{1}{\rho}\frac{d\rho}{dT}$; 10°C-70°C	8.6-9.1	7.1-7.4	7.7-8.1

*Private communication - K. Polmanteer

of viscous flow is reported to be a measure of the energy required to create a hole for a polymer flow segment. The E_{vis} values would indicate that as the pendant side group increases in size and/or polarity, the cohesive energy density increases. However, flow segment size might account for some of the differences.

An exceptionally low Glass Transition Temperature, T_g, for $(EtMeSiO)_n$ of -135°C in constrast to $(Me_2SiO)_n$ with a value of -123°C indicates greater flexibility at lower temperatures. Cohesive energies have been related to T_g(5). The low T_g observed for $(EtMeSiO)_n$ polymers would suggest a lower solubility parameter than for $(Me_2SiO)_n$. On the other hand, $(F_3CCH_2CH_2MeSiO)_n$ compositions are used commercially as oil and fuel resistant materials. Its higher T_g and E_{vis} are consistent with an increased solubility parameter.

The thermal expansion coefficient, $-\frac{1}{\rho}\frac{d\rho}{dT}$, was measured on a Mettler Parr Densimeter for $(Me_2SiO)_n$ and $(MeEtSiO)_n$ polymers and found to be respectively 9×10^{-4} $°C^{-1}$ and

7×10^{-4} °C^{-1} suggesting that the cohesive energy density might be different for the two polymers if the compressibility did not compensate for the difference(6). The obviously more polar polytrifluoropropylmethylsiloxane has a thermal expansion of 8×10^{-4} °C^{-1} intermediate to polyethylmethylsiloxane and polydimethylsiloxane.

III. EXPERIMENTAL

A. Swelling Studies

High molecular weight gums of $(Me_2SiO)_n$, $(EtMeSiO)_n$, or $(CF_3CH_2CH_2MeSiO)_n$ were degassed and pressed in 3" x 8" chases to form films. These films were irradiated with γ radiation at room temperature for a dosage of 7 megarads to yield lightly crosslinked elastomers. Samples (approximately 2cm x 2cm x 0.2cm) were cut from the elastomers and weighed. They were swollen with various solvents (or solvent combinations) in 4 oz., widemouthed jars and reweighed after 24 hrs. and 140 hrs. by removing the samples from the solvent, quickly removing excess solvent by blotting the sample between sheets of filter paper, and weighing the sample in a glass jar with a lid. The samples were then air dried in glass jars in the bottom of a forced-air hood for 3 days. The dried samples were weighed and the volumes of the samples determined via a simple specific gravity technique which measures the volume of water displaced by the sample in a picnometer jar. The % volume swell used in this study is defined by the equation below.

$$\%\ volume\ swell = \frac{volume\ of\ swollen\ sample}{volume\ of\ dried\ sample} \times 100\%$$

The swell data are shown in Figures 1 and 2 and Table 5.

B. Determination of Partial Solubility Parameters

The three parameter values for polymers are generally determined by assuming that all non-solvents fall outside an ellipsoidal region defined by the parameters and that all solvents fall inside.(7)

Using the data of Figures 1 and 2, the polymer ellipsoidal region of solubility was arbitrarily defined by surfaces of either 300% or 400% volume swell. Partial solubility parameters of these surface points were calculated and input to a computer program along with the normalized coefficients (γ_1, γ_2, γ_3) of the ellipsoid as 4, 1, 1 and an estimate of the center's coordinates. The initial estimate was made by observing the approximate center of two dimensional plots or with the aid of a computer locating the intersection of lines normal to combinations of three surface points. The program searches out the best possible solution of polymer δ_d, δ_p, δ_h, and radius R to the equation of the ellipsoid:

$$R^2 = \gamma_1 (\delta_{d_2} - \delta_d)^2 + \gamma_2 (\delta_{p_2} - \delta_p)^2 + \gamma_3 (\delta_{h_2} - \delta_h)^2$$

TABLE 5
Solvent welling of $(Me_2SiO)_n$ $(EtMeSiO)_n$ and $(CF_3CH_2CH_2MeSiO)_n$

Solvent	% Volume Swell $(Me_2SiO)_n$	$(EtMeSiO)_n$	$(CF_3CH_2CH_2MeSiO)_n$
Perfluoro-methylcyclohexane	104	102	105
Octamethylcyclotetrasiloxane	510	542	102
2,2,4-trimethylpentane	550	742	113
n- hexane	578	784	113
n- heptane	688	764	112
Methylcyclohexane	598	750	111
Cyclohexane	559	700	112
n- propylbenzene	406	520	115
Ethylbenzene	417	509	118
Toluene	425	670	124
o- xylene	389	512	119
Benzene	362	455	126
sec-butanol	157	163	105
n- butanol	127	122	103
n- propanol	117	111	103
Ethanol	108	104	103
Methanol	103	101	103
o- dichlorobenzene	181	218	114
Ethylene Dichloride	157	135	154
Cyclohexanone	125	122	445
Dioxane	144	125	140
Methyl Ethyl Ketone	155	135	860
Acetonitrile	103	101	160
Ethyl Acetate	281	155	580
Methyl i-butyl Ketone	323	351	954
Chloroform	514	706	141
Hexamethyldisiloxane	531	716	--
Di-isobutyl Ketone	393	518	630

The output includes an error for each surface point and a standard deviation of each of the partial solubility parameters and radius. The partial solubility parameters for the polymers are shown in Table 6.

IV. RESULTS AND DISCUSSION

The single-parameter solubility parameter can be determined via numerous approaches.(8) Table 7 summarizes single-parameter solubility parameter values for several $(RMeSiO)_n$ polymers determined by group additivity,(9) select solvent swelling,(10) or refractive index methods.(11) Group additivity values of δ are in excellent agreement for previously reported select solvent swelling values. The δ values for $(EtMeSiO)_n$ were found by select solvent swelling to be 7.5 $cal^{1/2}cm^{-3/2}$, the same as for $(Me_2SiO)_n$. The additivity method would suggest a slightly higer value. The dispersive partial solubility parameter as determined from refractive index remained fairly constant for the $(RMeSiO)_n$ polymers examined suggesting that the primary contribution may be from the Si-O-Si backbone structure. Variations in the total δ are brought about by side group interactions.

In order to ascertain structural contribution, the three-component partial parameters were determined. Values for the partial parameters of the solvents employed were all taken from reference(7).

Using the percentage volume swell described in reference (10), R. Gee of our laboratories has calculated the partial parameters for $(Me_2SiO)_n$ polymers to be $\delta_d = 7.6$ $cal^{1/2}cm^{-3/2}$; $\delta_p = 0.8$ $cal^{1/2}cm^{-3/2}$; $\delta_h = 1.7$ $cal^{1/2}cm^{-3/2}$ by the calculation procedure described in the experimental section and assuming a linear relationship between percent swell and solvent pair composition. This assumption requires ideal solvent-solvent mixtures. Our studies, Figures 1 and 2, indicate that a linear relationship is seldom encountered.

The partial parameter values determined for $(Me_2SiO)_n$ from the ideal solvent mixture assumption of Gee are quite close to the values determined from our non-linear mixture data. Because of the agreement of the linear interpolation and solvent mixture results and because of the rather enormous

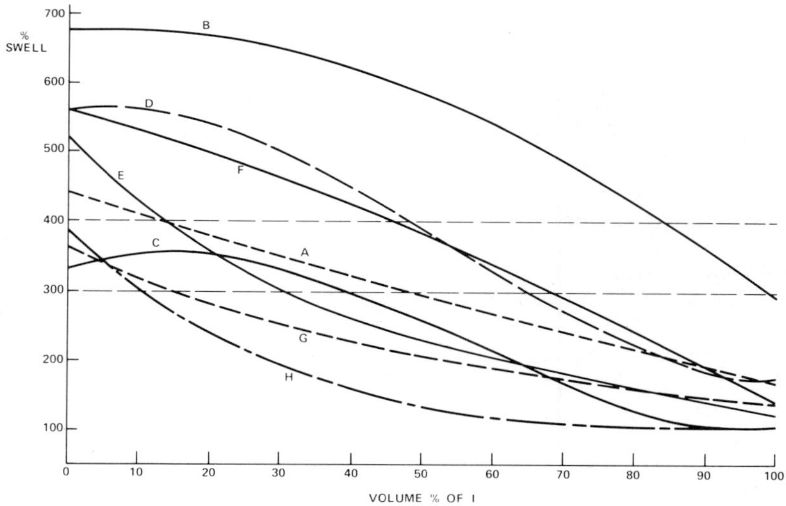

Fig. 1. Volume % swell of $(Me_2SiO)_n$ in solvent pairs I and II. A: I= o-dichlorobenzene, II= toluene; B: I= ethyl acetate, II= n-C_7H_{16}; C: I= CH_2Cl_2, II= o-xylene; D: I= MEK, II= cyclo-C_6H_{12}; E: I= dioxane, II= $CHCl_3$; F: I= sec-BuOH, II= methylcyclohexane; G: I= cyclohexanone, II= benzene; H: I= MeCN, II= di-isobutylketone

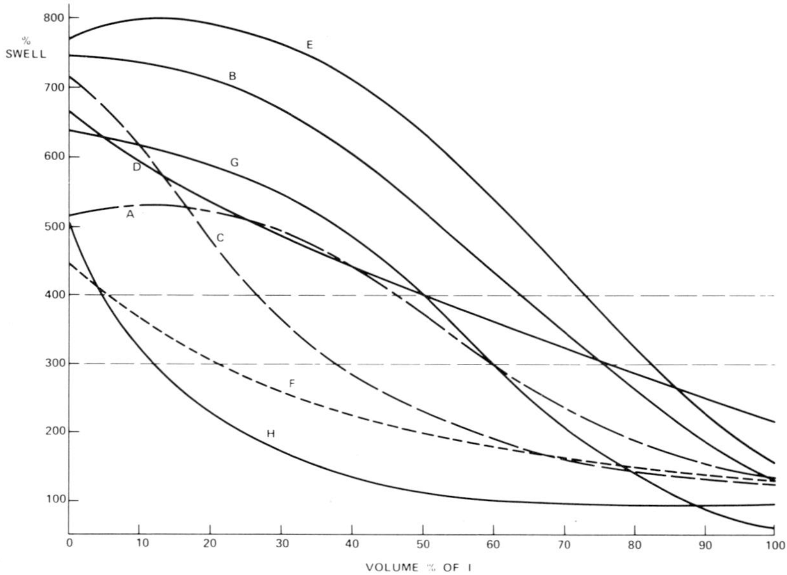

Fig. 2. Volume % swell of $(EtMeSiO)_n$ in solvent pairs I and II. A: I= CH_2Cl_2, II= o-xylene; B: I= MEK, II= cyclo C_6H_{12}; C: I= dioxane, II= $CHCl_3$; D: I= o-dichlorobenzene, II= toluene; E: I= ethyl acetate, II= n-C_7H_{16}; F: I= cyclohexanone, II= C_6H_6; G: I= sec-BuOH, II= methylcyclohexane; H: I= MeCN, II= di-isobutyl ketone

task of measuring swell in a series of solvent mixtures, partial parameters for polytrifluoropropylmethylsiloxane were determined by the linear interpolation method only. The arbitrary choice of swell boundry, 300% or 400%, seemed to have little effect upon the precision of the measurements reported in Table 6.

Polyethylmethylsiloxane and polydimethylsiloxane have very similar partial parameters,(Table 6). The polar component is nearly zero and not unexpected. The wide Si-O-Si bond angle(1) of 130° and low barrier of rotation(4) around this Si-O-Si bond contribute to give a low dipole moment to the Si-O-Si bond. The dipole for $Me_3SiOSiMe_3$ has been found to be 0.78D(12). In contrast, polytrifluoropropylmethylsiloxane has a high polar component arising from the very polar side group $CF_3CH_2CH_2$-. The Taft(13) substituent constant σ, which is indicative of the electronic effect of a substituent, is reported to be about 1.0(14). On the same scale, a methyl value is arbitrarily assigned a value of 0.0 and an ethyl group has a value of -0.1. Hydrogen bonding to the Si-O-Si bond is lower than to C-O-C bonds, but not negligible(15). Thus, the hydro-

TABLE 6
Partial Solubility Parameters for $(Me_2SiO)_n$ and $(EtMeSiO)_n$

Polymer	*% Swell of Ellipsoid Surface*	δ_d	δ_p	δ_h
$(Me_2SiO)_n$	300	7.8 ± 0.1	0.05 ± 0.2	2.3 ± 0.2
$(Me_2SiO)_n$	400	7.8 ± 0.1	-0.02 ± 0.2	1.7 ± 0.3
$(EtMeSiO)_n$	300	7.9 ± 0.2	0.7 ± 0.4	2.0 ± 0.4
$(EtMeSiO)_n$	400	8.0 ± 0.1	0.1 ± 0.2	1.9 ± 0.3
$(CF_3CH_2CH_2MeSiO)_n$	300	7.6 ± 0.6	3.5 ± 0.5	2.8 ± 0.9
$(CF_3CH_2CH_2MeSiO)_n$	400	8.0 ± 0.1	3.1 ± 0.1	2.5 ± 0.2

gen bonding component found is also very reasonable. Hydrogen bonding to a fluorine atom in $(CF_3CH_2CH_2MeSiO)_n$ would be expected to be low. Hydrogen bonding to C-F bonds is only about 2kcal(16). The larger size of silicone and the resulting more polarizable bonds contribute significantly to molar refraction (17) and thus, to the dispersion component. It is not surprising that the dispersive components are quite similar since all of the polymers in the series, $(RMeSiO)_n$ have the same number of Si-C and Si-O bonds. The dispersive component found for the polymers is in good agreement with the values calculated from the refractive index (Table 7). It is obvious from

TABLE 7
Solubility Parameters for Several $(RMeSiO)_n$ Polymers

R	$\delta^{a.}$	$\delta^{b.}$	$\delta^{d.}$
Me-	7.4cal$^{1/2}$cm$^{-3/2}$ $^{e.}$	7.5cal$^{1/2}$cm$^{-3/2}$	7.5cal$^{1/2}$cm$^{-3/2}$
Et-	7.7	7.5$^{c.}$	7.9
Ph-	9.7	9.6	7.8
$CF_3CH_2CH_2$-	9.0	9.0	7.1

a. Group additivity reference 9; b. Data from reference 10, select solvent swelling method; c. Select solvent swelling method - this study; d. From refractive index, reference 11; e. Multiply all values by 2.0455 to convert to SI units, MPa$^{1/2}$.

these studies that $(Me_2SiO)_n$ and $(EtMeSiO)_n$ have very similar polymer interactions and $(CF_3CH_2CH_2MeSiO)_n$ is unique because of the polar contribution of the trifluoropropyl side group.

IV. REFERENCES

1. Owens, M. J. and Evans, John I., Br. Polymer J., 7, 235 (1975).
2. Eaborn, C., "Organosilicon Compounds," Butterworths Scientific Publications, London (1960) p. 89.
3. Ebsworth, E. A. V., in "The Bond to Carbon, Part I," ed. MacDiarmid, A. G., Marcel Dekker, Inc., New York (1968), p. 17.

4. Scott, D. W., Messerly, J. F., Todd, S. S., Guthrie, G. R., Hossenlopp, I. A., Moore, R. T., Osborne, Ann, Berg, W. T., and McCullough, J. P. J. Chem. Phys. 65, 1320-26 (1961).
5. Hayes, R. B., J. Poly. Sci., 5, 318 (1961).
6. Hildebrand, J. H. and Scott, R. L., "The Solubility of Nonelectrolytes," Rheinhold Publishing Corp., New York (1960).
7. Hansen, C. and Beerbower, A., in Kirk-Othmer, "Encyclopedia of Chemical Technology", Suppl. Vol. 2nd Ed., A Standon, Ed., 1971 p. 889.
8. Barton, A. F. M., Chem. Rev., 75, (6) 731 (1975).
9. Fedors, R. F., Poly. Eng. and Sci., 14, 147 (1974).
10. Yerrick, K. B. and Beck, H. N., Rubber Chem. and Tech., 37, 261 (1964).
11. Keller, R. A., Kangor, B. L. and Snyder, L. R., Gas Chromatogr. Proc. Int. Symp. (Eur.), 8, 125 (1971).
12. Feiser, H., Eagle, M.V., Speier, J. L., J. Amer. Chem. Soc., 75, 2824 (1953).
13. Taft, Jr., R. W., in "Steric Effects in Organic Chemistry," Newman, M. S., Ed., J. Wiley and Sons, Inc., New York, N. Y., 1956, p. 556.
14. Lipowitz, J., J. Amer. Chem. Soc., 94, 1582 (1972).
15. West, R., Whatley, L. S. and Lake, K. J., J. Amer. Chem. Soc., 83, 761 (1961).
16. Jones, D. A. K. and Watkinson, J. G., J. Chem. Soc., (1964) 2366.
17. Warrick, E., J. Amer. Chem. Soc., 68, 2455 (1946).

THE EFFECT OF STRUCTURAL MODIFICATIONS ON SOLVENT INTERACTIONS WITH SILOXANE POLYMERS

D. C. Bonner, K. C. Chen, and H. Rosenberg
Texas A&M University, Texas Tech University, and Air Force Materials Laboratory

I. INTRODUCTION

Since the pioneering work of Smidsrød and Guillet(1), numerous investigators have made use of gas-liquid chromatography (GLC) to determine infinite-dilution activity coefficients and partial molar enthalpies characteristic of the interactions of small amounts of solvents with polymers. For a review see the article by Bonner(2).

In the present study, we investigated the sensitivity of GLC data to structural modifications in a series alkyl- and fluoroalkyl-substituted siloxane polymers. The structures of the polymers studied, including six recently synthesized by Rosenberg and Choe(3), are given in Figure 1. While polymers VI and VII

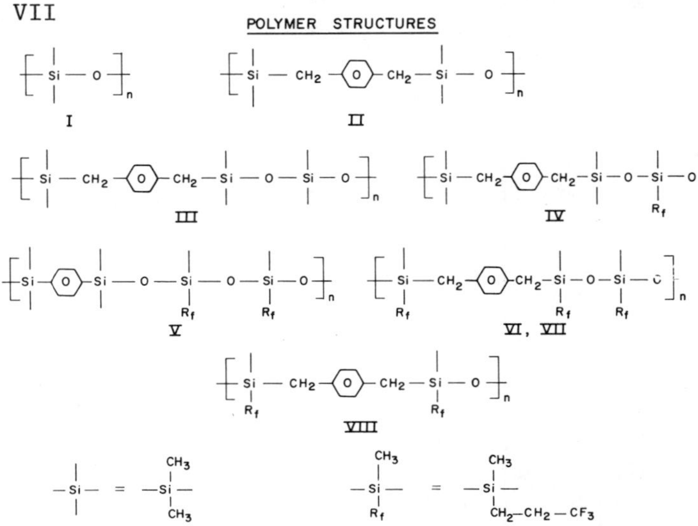

Fig. 1. Polymer structures.

have nominally the same structure, polymer VI contains 3% vinyl groups as copolymer.

II. EXPERIMENTAL

The gas-liquid chromatograph used in this investigation was a Varian Model 920. The experimental apparatus has been described earlier by Cheng and Bonner(4).

The siloxane polymers were coated on Fluoropak 80 (80/100 mesh) from acetone solution. Column loadings are shown in Table I. Early GLC studies with siloxane polymers demonstrated (5,6)

TABLE 1.
Column loadings

Column	g. Polymer	g. Fluoropak	% Loading
I	0.381	2.817	11.9
II	0.0602	0.988	5.74
III	0.0727	0.915	7.36
IV	0.0956	1.084	8.10
V	0.0981	1.170	7.74
VI	0.0205	0.586	3.38
VII	0.137	1.373	9.07
VIII	0.0632	1.064	5.61

that column loading, helium flow rate, and solute sample size have little effect on the thermodynamic quantities derived from GLC measurements.

The specific retention volume corrected to $0\,°C\,(V_g^{\circ})$ was computed from the equation(4)

$$V_g^{\circ} = Q(t_g - t_r)(273.15/T)(f_p/m_2) \tag{1}$$

where Q = carrier gas flow rate measured at the column outlet
m_2 = mass of polymer coated on the column
t_g = solute retention time (to peak maximum)
t_r = air retention time (to peak maximum)
T = column temperature

There is a pressure drop through the column, so the factor f_p is used to correct V_g° to average column pressure. The pressure correction, f_p, is given(4) by

$$f_p = (3/2)[(P_i/P_o)^2-1]/[(P_i/P_o)^3-1] \tag{2}$$

Retention volumes for n-hexane, n-heptane, benzene, chloroform, ethanol, n-propanol, and acetone were obtained for polydimethylsiloxane at 55°C and for seven other siloxane polymers at 55°C, 70°C, and 85°C.

III. POLYMER SAMPLES

Polymer samples used in this study were synthesized by Rosenberg and Choe(3). Molecular weights of the samples range from 10,000 to 50,000 g/gmole. The glass-transition temperatures for polymers II through VIII were determined by differential scanning calorimetry (DSC) and found to be -41°C, -62°C, -52°C, -68°C, -35°C, -33.5°C, and -19°C, respectively.

IV. ANALYSIS OF DATA

Retention volumes at more than one temperature provide sufficient information to calculate infinite-dilution activity coefficients of solutes in a polymer as well as solute partial molar enthalpies. At modest pressures, the weight-fraction activity coefficient of a solute at infinite dilution is given by(4)

$$\Omega_1^\infty = \lim_{w_1 \to o} (a_1/w_1)$$

$$= (273.2R)/(P_1^s V_g^\circ M_1)\exp[(-P_1^s B_{11})/(RT)] \tag{3}$$

where a_1 is solute activity, w_1 is solute weight fraction in solution, R is the gas constant, P_1^s is the saturation vapor pressure of the solute at column temperature T, M_1 is solute molecular weight, and B_{11} is the solute second virial coefficient at T.

If V_g° values are available at more than one temperature, the partial molar enthalpy can be calculated:

$$\left[\frac{\partial \ln \Omega_1^\infty}{\partial\,(1/T)}\right]_p = \frac{\bar{h}_1^\infty}{R} \tag{4}$$

where $\bar{h}_1^\infty$ is the partial molar enthalpy of the solute at infinite dilution.

Solute activity coefficients of n-hexane and chloroform for polymers II-VIII are plotted in Figures 2 and 3, respectively. Activity coefficients at 55°C are presented in Table 2.

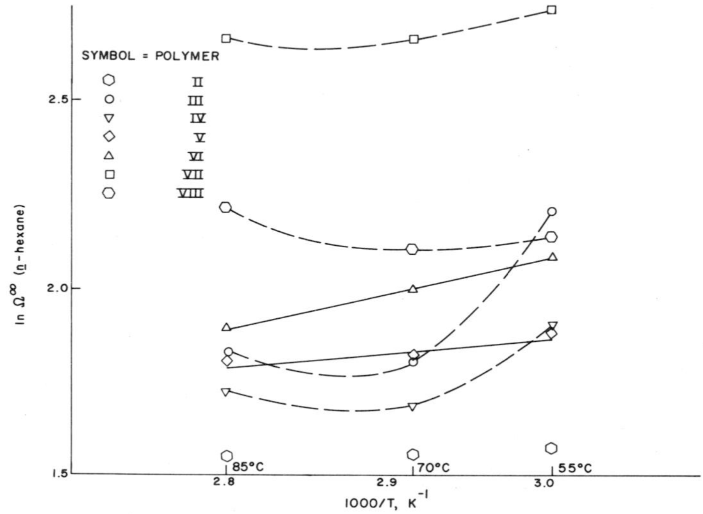

Fig. 2. Hexane activity coefficients in polymers II-VIII as functions of temperature.

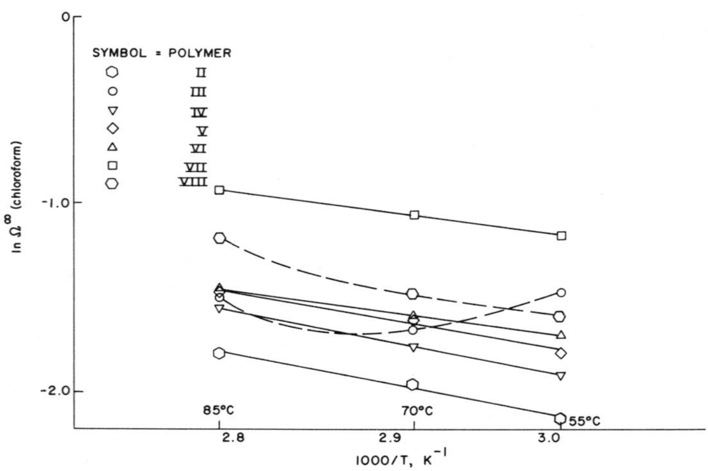

Fig. 3. Chloroform activity coefficients in polymers II-VIII as functions of temperature.

Partial molar enthalpies at 55°C are given in Table 3. The partial molar enthalpies were calculated from the slope of a least-squares straight line through data plotted as $ln\ \Omega_1^{\infty}$ versus $(1/T)$ for each solute/polymer combination. It should be noted, however, that some of the $ln\ \Omega_1$ versus $(1/T)$ plots are nonlinear. The causes and consequences of nonlinear behavior will be discussed below.

Retention volume values are estimated to be precise to ± 3%, activity coefficients to ± 5%, and partial molar enthalpies to ± 10% (for linear data).

TABLE 2.
Infinite-dilution activity coefficients of solvents in polymers at 55°C.

Solvent[a] / Polymer	C_6	C_7	*Bz*	$CHCl_3$	*EtOH*	*PrOH*	*Acet*
I	6.40	6.23	6.57	—	84.7	60.0	20.3
II	4.81	4.48	3.41	0.117	32.2	47.8	9.21
III	9.11	7.77	6.33	0.229	61.1	104.	20.1
IV	6.70	5.12	3.86	0.147	50.8	34.9	6.84
V	6.32	6.30	4.41	0.167	32.6	44.8	5.48
VI	8.06	8.41	4.23	0.183	31.1	42.5	4.30
VII	15.6	14.2	6.63	0.294	47.9	67.9	6.19
VIII	9.74	8.73	4.42	0.202	53.3	36.8	5.08

a. C_6 = n-hexane, C_7 = n-heptane, *Bz* = benzene, $CHCl_3$ = chloroform, *EtOH* = ethanol, *PrOH* = n-propanol, *Acet* = acetone.

In order to put our results in a more usual form, we have assumed that the Flory-Huggins(7,8) theory of polymer solutions is approximately correct for these solutions. Applying this assumption, one finds that

$$\bar{h}_1^{\infty} = RT\,\chi \tag{5}$$

where χ is the Flory-Huggins interaction parameter. Values of χ less than 0.5 indicate complete miscibility, while negative χ values are characteristic of extremely strong association in solution . Values of χ at 55°C computed on the basis of equation(5) are presented in Table 4.

V. DISCUSSION OF RESULTS

Referring to Figure 2, one can see that at 55°C hexane interactions with polymers II-VIII are, in order of decreasing χ value, II>V>I>IV>VI>III>VIII>VII. At 85°C, however, the order is II>IV>V>III>VI>VIII>VII. The reason for the change in order between 55°C and 85°C is that the plots of

$ln\Omega^{\infty}$ versus $1/T$ for polymers VII, VIII, III, and IV are non-linear.

TABLE 3.
Partial molar enthalpies of solvents at infinite dilution, in kcal/gmole.

Solvent[a] Polymer	C_6	C_7	*Bz*	$CHCl_3$	*EtOH*	*PrOH*	*Acet*
I	—	—	—	—	—	—	—
II	0.208	0.179	0.246	-2.68	3.79	3.63	0.944
III	2.95	3.08	2.95	0.207	6.16	7.07	4.53
IV	1.40	-0.156	-2.31	-2.77	4.54	3.57	-0.101
V	0.270	0.332	0.174	-2.53	3.66	3.57	0.0280
VI	2.27	1.81	0.0247	-1.88	3.99	3.82	0.214
VII	0.543	1.05	0.0776	-1.11	3.25	3.43	-0.0724
VIII	0.491	-0.218	-1.15	-3.26	3.30	2.75	-0.0814

a. C_6 = n-hexane, C_7 = n-heptane, *Bz* = benzene, $CHCl_3$ = chloroform, *EtOH* = ethanol, *PrOH* = n-propanol, *Acet* = acetone

TABLE 4.
Flory-Huggins Interaction Parameters of Solutes at Infinite Dilution.

Solvent[a] Polymer	C_6	C_7	*Bz*	$CHCl_3$	*EtOH*	*PrOH*	*Acet*
I	—	—	—	—	—	—	—
II	0.319	0.274	0.377	-4.11	5.81	5.57	1.45
III	4.52	4.72	4.52	0.317	9.44	10.8	6.95
IV	2.15	-0.239	-3.54	-4.25	6.96	5.47	-0.155
V	0.414	0.509	0.267	-0.388	5.61	5.86	0.0430
VI	3.48	2.78	0.038	-2.88	6.12	5.26	0.328
VII	0.833	1.61	0.119	-1.70	4.98	5.26	-0.111
VIII	0.753	-0.334	-1.76	-5.00	5.06	4.22	-1.25

a. C_6 = n-hexane, C_7 = n-heptane, *Bz* = benzene, $CHCl_3$ = chloroform, *EtOH* = ethanol, *PrOH* = n-propanol, *Acet* = acetone

It is generally agreed that non-linearity of the type shown in Figure 2 is caused by a physical change in polymer morphology (melting; glass transition; possible a third-order transition) or by errors in data. We do not know the reasons for the curvature of data for polymers VII and IV, but polymer III has a melting point at 61°C as determined by DSC, and DSC results for polymer VIII show a very weak transition between 40°C and 60°C.

Activity coefficients for chloroform, shown in Figure 3 reveal a similar pattern, although the activity coefficients for chloroform are quite small. Chloroform is therefore a good solvent for the polymers studied(2).

Only two polymers, III and VIII, show curvature with chloroform. Polymers III and VIII are the only two polymers for which explantions of data curvature are available. Although chloroform is a much better solvent for polysiloxanes than is _n_-hexane, the ranking of polymer susceptibility to dissolution is essentially the same as that in hexane.

Considering polymer/solvent interactions at 85°C, which appears to be a temperature greater than those of possible transitions, a clear picture of the cause of solvent ranking becomes apparent. Polymer IV is less soluble than polymer II because of the addition to each repeat unit of a siloxane moiety containing a trifluoropropyl substituent. Polymer V becomes even less soluble than polymer IV because of the replacement of a second methyl group by a trifluoropropyl group. Polymer III is apparently less soluble than polymer II because the number of siloxane units is larger by one, reducing the effect of the xylylene moiety. The xylylene moiety tends to enhance polymer solubility, as can be seen by comparison of activity coefficients of polymers I and II, shown in Table 1.

Polymer VI (which contains 3% vinyl groups) is less soluble than polymer V because of the presence of a third trifluoropropyl group. Polymer VIII is yet less soluble, perhaps because of steric hinderance of solvent by trifluoropropyl groups in a short repeat unit. The short repeat unit also results in a slightly less flexible chain.

Finally, polymer VII (without any vinyl groups) is the least soluble because of a high density of trifluoropropyl groups working against solvent interaction with the xylylene moiety, and there are no vinyl groups present to improve solubility.

Partial molar enthalpies at infinite dilution relate essentially the same ranking of solvents in the polysiloxanes as do activity coefficients. It is noteworthy that all partial molar enthalpies of chloroform in polymers II-VIII are negative, indicating an exothermic association in solution. Chloroform probably acts as a Lewis acid, which would tend to associate exothermically.

Acetone also may exhibit exothermic interactions with

polymers IV, VII, and VIII. However, the partial molar enthalpy of acetone in polymers IV and VII is sufficiently close to zero that, within experimental error, the interactions may be considered athermal. The interaction between acetone and polymer VIII is exothermic and of sufficient magnitude that the partial molar enthalpy could not be considered zero within experimental error.

Other exothermic interactions are exhibited by *n*-heptane in polymers IV and VIII and by benzene in columns IV and VIII. Again, the magnitude of the partial molar enthalpies for benzene and *n*-heptane interactions is sufficiently small that within experimental error, they may be considered athermal.

VI. CONCLUSIONS

We have demonstrated that gas-liquid chromatography is very sensitive to structural modifications of polysiloxanes. Also, we have shown that gas-liquid chromatography can be used to obtain valuable information not only concerning solute solubility in polysiloxanes but also concerning the enthalpy of interaction between solutes and polymers.

Finally, gas-liquid chromatography can be used to detect, in a qualitative way, changes in polymer morphology with temperature.

VII. REFERENCES

1. Smidsrød, O., and Guillet, J. E., *Macromolecules*, 2, 272 (1969).
2. Bonner, D. C., *J. Macromol. Sci.- Revs. Macromol. Chem.*, C13, 263 (1975).
3. Rosenberg, H., and Choe, E. W., AFML-TR-75-182, Part II, June, 1976; Abstr. of Papers, Div. Fluorine Chem., 172nd National Mtg., Amer, Chem. Soc., San Francisco, Ca., August, 1976.
4. Cheng, Y. L., and Bonner, D. C., *Macromolecules*, 7, 687 (1974).
5. Summers, W. R., Tewari, Y. B., and Schreiber, H. P., *Macromolecules*, 5, 12 (1972).
6. Lichtenthaler, R. N., Newman, R. D., and Prausnitz, J. M., *Macromolecules*, 6, 650 (1973).
7. Flory, P. J., *J. Chem. Phys.*, 9, 660 (1941).
8. Huggins, M. L., *J. Chem. Phys.*, 9, 440 (1941).
9. Guillet, J. E., *Adv. Gas Chromatogr.*, 11, 187 (1973).

SOLUTIONS OF CELLULOSE IN ORGANIC SOLVENTS

Raymond B. Seymour and Earl J. Johnson
University of Southern Mississippi

I. INTRODUCTION

Cellulose is a naturally occurring colorless, linear macromolecular carbohydrate consisting of anhydroglucose units joined together by beta acetal linkages. The number of anhydroglucose units or degree of polymerization (D.P.) in naturally occurring cellulose is about 3000 and about 200 in alpha cellulose.

While cellulose occurs as an essentially pure polymer in cotton, it is admixed with lignin, hemi-celluloses and minerals in wood. These noncellulosic constituents are removed mechanically or chemically in the manufacture of paper. Attempts were made to preferentially dissolve the cellulosic constituents in this investigation.

Cellulose is a highly crystalline polymer in which the linear polymeric chains are held together as micelles by relatively strong intermolecular hydrogen bonds. The strength of these bonds is reduced in the presence of nitric acid, sodium hydroxide, zinc chloride, cupric ammonium hydroxide, tetraalkyl ammonium hydroxide, or sulfuric acid.

The solubility of the cellulose nitrates produced by the reaction of cellulose and nitric acid is dependent on the degree of substitution (D.S.). In his serendipitous discovery of collodion, Menard found that cellulose nitrate with a D.S. of about 2 was soluble in an equimolar mixture of ethanol and ethyl ether.

The soda cellulose produced when cellulose is immersed in aqueous alkaline solutions is used to produce mercerized cotton and cellulose ethers. Carboxymethylcellulose (CMC), methylcellulose and hydroxyethylcellulose which are soluble in water are obtained by the reaction of soda cellulose with chloroacetic acid, dimethyl sulfate, and ethylene oxide, respectively. Cellulose xanthate (viscose) is obtained by the reaction of carbon disulfide with soda cellulose. Rayon and cellophane are produced by the regeneration of cellulose which occurs when the soluble viscose solution is precipitated in water.

Vulcanized fiber is produced by passing paper through warm aqueous zinc chloride and washing and drying the dark colored produce. Rayon has been produced by dissolving cellulose in a solution of copper hydroxide in aqueous ammonia (Schweitzer's reagent) and precipitating this solution in water. Parchment paper has been prepared by washing and drying paper that has been dipped in 80 percent sulfuric acid.

Cellulose is not soluble in the following polar organic solvents with solubility parameters (δ) as shown: acetone (δ = 10.0H), dimethylacetamide (δ = 10.0H), acrylonitrile (δ = 10.5H), hexamethylphosphoramide (δ = 10.5H), acetonitrile (δ = 11.94H), dimethylformamide (δ = 12.1H), nitromethane (δ = 12.7H), sulfolane (δ = 12.5H) or dimethyl sulfoxide (DMSO) (δ = 12.0H). However, dry cellulose is soluble in DMSO, but not in other polar solvents in the presence of formaldehyde (1) (2).

II. EXPERIMENTAL

A clear viscous solution was obtained when 1 g of dry cellulose in 100 ml of DMSO was heated with occasional agitation at 70° in the presence of 2.0g paraformaldehyde or 0.5g formaldehyde for twelve hours. After the removal of excess formaldehyde by further heating in an open vessel, this cellulose solution was mixed in various ratios with solutions of polyvinyl alcohol, polyacryonitrile and polyvinyl pyrrolidone in DMSO.

The cellulose in the DMSO solution was also esterified at room temperature by the addition of 1.0 ml of pyridine and varying amounts of acid anhydride to 10 g of a 0.75 percent solution of cellulose. The solution of the ester was filtered through a coarse sintered glass funnel, extracted with water and dried in a vacuum desiccator.

All products were characterized by proton magnetic resonance spectrometry using a Varian T-60 spectrometer, by infrared spectrophotometry using a Perkin Elmer 700 spectrometer, by ultraviolet spectrophotometry using a Beckman DB-CT dual beam spectrometry and by differential scanning calorimetry using a Perkin Elmer 990 thermal analyzer.

III. DISCUSSION

Since cellulose dissolves in DMSO in the presence of formaldehyde, it is assumed that methylene diol must be present for solubility to occur (3). It is also assumed that a transcient six membered complex which is formed between DMSO and the hydroxyl groups in methylcellulose interferes with the normal intermolecular hydrogen bonding in cellulose (4).

The viscosity of these solutions of high molecular weight cellulose at 25° increased from 2.6 cps for a 0.5 percent solution to 6.0 cps for a 2.5 percent solution. A gel was produced when attempts were made to produce more concentrated solutions.

While cellulose solutions in aqueous solvents are compatible with aqueous solutions of poly vinyl alcohol, starch and polyvinylpyrrolidone, it is not possible to prepare solutions of mixtures of cellulose and polyacrylonitrile in aqueous media. However, blends of all possible ratios of cellulose in DMSO were prepared with DMSO solutions of poly vinyl alcohol, polyvinylpyrrolidone and polyacrylonitrile.

The viscosity of the blends of cellulose and polyacrylonitrile increased as the concentration of the latter component was increased. Strong films and fibers were obtained when the polymer blends were regenerated by evaporation or by precipitation in water. Thus, it was possible to reduce the water absorption of cellulose or to increase the water absorption of polyacrylonitrile by the selection of appropriate ratios of blends of the two component polymers.

While no noticeable esterification occurred when anhydrous carboxylic acids were added to solutions of cellulose in DMSO, esters were produced when acid anhydrides were added to these solutions in the presence of pyridine at room temperature. Most esters darkened as the esterification time increased.

However, essentially colorless thermoplastic esters with a D.S. of less than 2 were obtained with short reaction time (15 - 30 min.). The esters were recovered by precipitating in water. Esters with D.S. values from

0.1 to 2.0 were obtained by esterification of the cellulose solution in DMSO with acetic anhydride, butyric anhydride, methacrylic anhydride, phthalic anhydride, maleic anhydride and succinic anhydride. The succinates were colorless and water soluble.

Attempts to produce alkylethers of cellulose were not successful but hydroxyethylcellulose was obtained when ethylene oxide was added to solutions of cellulose in DMSO.

IV. CONCLUSIONS

While cellulose is insoluble in most organic solvents, it does dissolve in DMSO in the presence of anhydrous formaldehyde. These solutions are compatible in all proportions with solutions of polyacrylonitride in DMSO. Esters with a D.S. of 0.1 to 2.0 are produced when organic anhydrides are added to the cellulose - DMSO = solutions in the presence of pyrridine.

V. REFERENCES

1. Johnson, D.C., Nicholson, M.D. and Haigh, F.C., 8th Cellulose Conference, Syracuse, New York, June 1975.

2. Seymour, R.B., and Johnson, E.L., Journal of Applied Polymer Science 20,000 (1976).

3. Chapman, A.L. and King, R.W., J. Am. Chem. Soc., 86,1256, 4968 (1964).

4. Jacob, S.W., Rosenbaum, E.E. and Wood, D.C., "Dimethylsulfoxide", Marcel Dekker, Inc., New York, 1971.

VI. ACKNOWLEDGEMENTS

The partial support of the Robert A. Welch Foundation is gratefully acknowledged.

THE EFFECTS OF ACRYLAMIDE GRAFT COPOLYMERIZATION ON THE SOLUBILITY AND VISCOSITY OF SEVERAL POLYSACCHARIDES

Donald R. Owen
University of Utah

Tomas C. Shen
University of Southern Mississippi

I. INTRODUCTION

Graft copolymers are becoming more important as an alternative to totally new linear polymers or copolymers. The ability to synthesize a polymer with unique solubility and rheological properties from readily available and reasonably inexpensive monomers is very attractive industrially. However, graft copolymers are inherently difficult to characterize. Therefore, relatively little exact information is available, even though many graft copolymers have been reported in the literature.

Theoretically a block or graft copolymer should offer clear advantages over a physical blend of the homopolymers when a high degree of incompatibility exists between the components. Also, one should be able to mate the physical characteristics of the backbone with the solubility properties of the side chain or terminal chains. The actual properties of a graft or block copolymer will, of course, vary according to the number and length of side or terminal chains. For a graft copolymer, as the segmental length decreases, the physical properties of the graft copolymer will approach those of a random copolymer. However, assuming that the graft segments are of sufficient length, the graft segments will markedly influence the solubility of the resulting copolymer. Depending on the relative solubility parameters of the particular components of the copolymer, the resulting copolymer will exhibit the solubility properties of each homopolymer simultaneously, or some intermediate solubility.

For this reason, it is most important to control the graft frequency and graft segment length in order to achieve a particular series of characteristics.

The initial part of this study deals with such an optimization of grafting acrylamide onto dextran, amylopectin and amylose in order to produce copolymers with unique solubility and rheological properties.

Numerous methods have been reported in the literature for initiating graft copolymerization. Reviews by Aggarwal (1) and Ceresa (2) serve to demonstrate the numerous avenues which are open to the polymer chemist. When polyol type polymers, such as the polysaccharides, are the substrates, the cerium (IV) ion initiation method has been used extensively (3-10) and was chosen as the method of initiation for this study because of its simplicity and controllability.

Model compound studies using cerium ion oxidation of monohydric alcohols and diols have given support to the postulated mechanism that the C2-C3 diol and the C6 hydroxyl of the anhydro-D-glucose unit are the preferred sites for free radical generation (3-7). However, other workers (8-10) have suggested that the terminal hemi-acetal groups are the predominant sites of free radical formation.

Evidence for the formation of stable coordination complexes has been obtained by kinetic and spectrometric methods for the cerium (IV) oxidations of numerous compounds in both perchloric and nitric acids (6,7,9-15). Complex formation in cerium (IV) oxidations in sulfuric acid solutions has been detected only in a few studies (16-20). Reactions in sulfuric acid have been reported to follow second order kinetics (3,6,13,19-21). Muhammad and Rao (21) have suggested a direct oxidation mechanism for oxidation in sulfuric acid. Hintz and Johnson (6) point out the possibility that the oxidation proceeds through an intermediate complex where the equilibrium constant formation is small.

The complex formation between cerium (IV) and a diol may be either a chelated or an acyclic species (22-25). The comparison of oxidations of monohydric alcohols (6,12,14,25) which cannot form chelate intermediates and diols (6,13,25) by cerium ion in 1.0 M perchloric acid show that the equilibrium constants for complex formation are significantly larger for the diol. This increase is attributed to stable chelate which forms in the cerium (IV) oxidation of the diol structure. When cyclohexanediols are compared with their monomethyl ethers (e.g. substitution of a methyl group for the hydroxyl hydrogen, which should retard chelate formation) the overall rates of oxidation were about equal. This apparently was due to compensating differences in equilibrium constants for the complex formation and the disproportionation rate constants for the two complexes.

Amis (26) reported the rate of oxidation of sugars by cerium (IV) in aqueous sulfuric acid seemed dependent on the free aldehyde concentration and the diol configuration. Based on these results, he suggested the following mechanism.

$$Ce(SO_4)_2 + HSO_4^- \overset{K_1}{\rightleftharpoons} Ce(SO_4)_3^{-2} + H^+$$

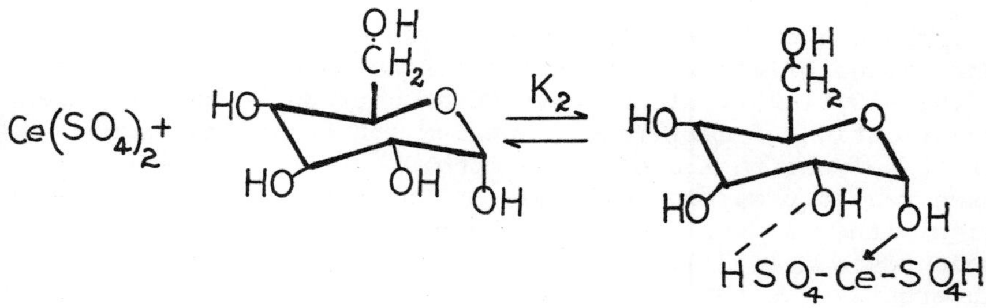

(Complex)

$$\text{Complex} \xrightarrow{K_1} \text{free radical} + Ce(III) + H^+$$

A similar mechanism in perchloric acid aqueous systems was suggested by Pottenger (27):

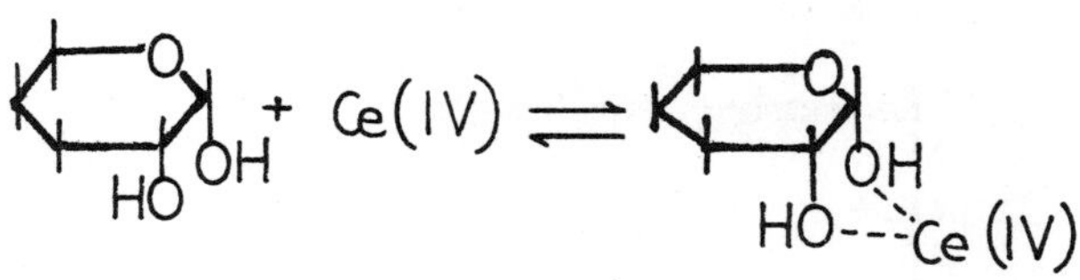

$$\text{Complex} \xrightarrow[\text{slow}]{K_1'} \text{free radical}$$

or

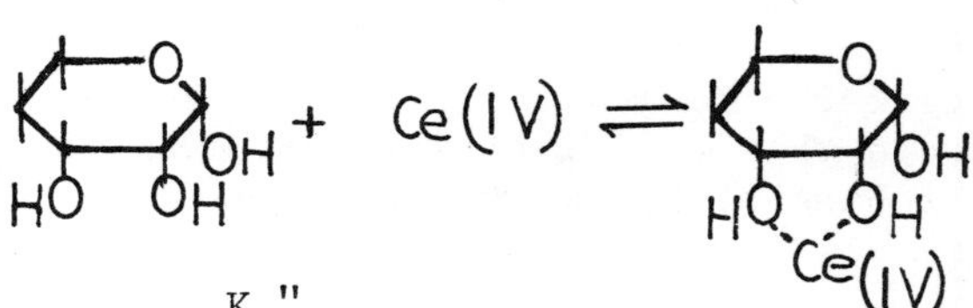

$$\text{Complex} \xrightarrow[\text{slow}]{K_1''} \text{free radical}$$

A comparative study of the cerium (IV) oxidation of ribose in aqueous solution containing approximately 10% free aldehyde groups and glucose containing 0.012% free aldehydic groups showed no significant rate difference. From this data Pottenger suggested the mechanism involves the pyranose form of the glucose molecule. The radical formation at C1 (as shown in the previous scheme) might well be preferred because of the greater resonance stabilization expected for such a radical. Hintz (6) compared the relative rates of cerium (IV) oxidation of transcyclohexanediol, cyclohexane methanol and tetrahydropran-2-methanol. From this study Hintz predicted that the C2-C3 diol structure in cellulose would be oxidized six times faster than the C6 hydroxyl.

The use of cerium (IV) alcohol redox systems for chemical initiation of vinyl polymerizations has been widely employed (3,4,8-10,21-29). There have been numerous studies on the modification of cellulose by cerium ion redox initiated graft polymerization of vinyl monomers. The formation of free radicals on the cellulose by the cerium ion redox reaction with a hydroxyl moiety has been confirmed by electron spin resonance (31). The mechanism by which cerium ion interacts with the substrate to form a free radical is beleived to involve the formation of a coordination complex between the oxidant and the hydroxyl groups of the cellulose. The cerium ion cellulose complex then disproportionates, forming a free radical on the cellulose chain as shown below.

$$\begin{array}{ll} \text{initiation} & Ce^{+4} + G \rightleftarrows \text{complex} \\ & \text{Complex} \longrightarrow \text{radical} \\ & \text{Radical} + M \longrightarrow M\cdot \\ \text{propagation} & M\cdot + M \longrightarrow M_n\cdot \\ \text{termination} & M_n\cdot + M_n\cdot \longrightarrow M_{2n} \\ & M_n\cdot + M_n\cdot \longrightarrow {}^{\sim}C{=}C + M_n \\ & M_n\cdot + Ce^{+4} \longrightarrow M_n + Ce^{+3} + H^+ \end{array} \qquad (4)$$

Young (10) studied the dextran acrylamide grafting reaction and concluded that the rapid observed rate of polymerization was due to the high reactivity of the cerium ion with the terminal cis-glycol groups.

Mino (30) investigated the polymerization of acrylamide which was initiated by ceric nitrate and 3-chloropropanol and reported that the rate of oxidation of the alcohol was decreased by the addition of the acrylamide. Presumably this occurs because of the complexing of acrylamide with the ceric ion. It was also reported that at constant pH and nitrate concentration, the rate of polymerization was independent of the initial cerium (IV) concentration. Kurosaki (4) used model compounds to study the dependency of acrylamide polymerization on cerium (IV) ion concentration. He determined that glucose which contains a hemiacetal unit was the most effective reducing agent and the structure of graft copolymer synthesized using a polyglucose type substrate was a mixture of block and graft type structures. In another acrylamide grafting study Takahashi (32) and Mino (30) investigated the cerium (IV) ion acrylamide complex formation by IR, NMR and x-ray. They determined that the ceric coordinated vinyl monomer system was capable of initiating the homopolymerization reaction under certain conditions.

Thus, our initial investigation centered on determining those conditions where homopolymerization was not ellicited and subsequently empolying these conditions for the graft copolymerization study.

II. MATERIALS

The acrylamide monomer was obtained as reagent grade from Eastman Chemical Corporation and purified by recrystalization from acetone. The amylopectin was obtained from Nutrition Biochemical Corporation as a purified grade with an intrinsic viscosity, as determined in our laboratory, of 1.17 and used as such. The dextran was also obtained from Nutrition Biochemical Corporation, with a reported molecular weight of 5-40 million and an intrinsic viscosity as determined in our laboratory of 1.5. The amylose was purchased from ICN pharmaceuticals, Inc. with a reported molecular weight of 300 glucose units, or 48,600.

The ceric ammonium nitrate was purchased from Pfaltz and Bauer Inc. and used without further purification. Sodium acrylate was synthesized and purified via washing with acetone in our laboratory according to the following procedure:

122.4 g (1.7 mole) of acrylic acid in 250 ml of acetone was p ured into a 1 liter flask held at 0 C. A solution of 68 g (1.7 mole) sodium hydroxide in 200 ml of distilled water was added slowly. The mixture was stirred for 2 hours. The precipitate was filtered and washed with acetone. A 74.2% yield of sodium acrylate was obtained after vacuum drying.

III. EXPERIMENTAL

A. Determining Conditions for Acrylamide Homopolymerization

Varying concentrations of acrylamide and nitric acid were placed in 25 ml reaction vessels with a Teflon coated rubber septum. After 30 minutes of purging with zero grade nitrogen the cerium (IV) ion solution was injected and the reaction, if any, allowed to continue for 24 hours.

B. Graft Copolymerization Condition Study

Various ratios of amylopectin, amylose or dextran to acrylamide, in measurements of 20 to 40 ml, were degassed using zero grade nitrogen for one hour in 50 ml septumed reaction vessels. The concentration of the substrate was varied from 0.0309 M to 0.0655 M of glucose units. Acrylamide concentrations were varied from 0.0585 M to 2.25 M and nitric acid concentrations from 0.045 M to 0.18 M. The cericum (IV) ion concentration was varied from 0.0001 M to 0.008 M. At 25° C the nitric acid and cerium (IV) ion solutions were injected into the reaction vessel. The reaction was allowed to continue for 1-3 hours, followed by termination via bubbling oxygen into the reaction solution. The solution was then subjected to dialyze for three days and freeze drying, thus isolating the polymer.

C. Synthesis of Graft Copolymers

The various concentrations of components are listed in Table I. The reaction procedure was the same as Section B. After terminating the reaction the mixture was dialyzed using cellulose acetate membrane tubing with a molecular weight cut off at approximately 3500. After 3 days of dialysis the solution was fractionated by the addition of a given amount of acetone (200 ml ~ 400 ml). The precipitated polymer solution was heated to 43° C to redissolve the precipitate. After standing for 24 hours at 25° C, the polymer gel was separated by decanting and freeze dried. The dried polymer was then dissolved in distilled water and filtered through a 40-60 millimicron sintered glass filter and subsequently freeze dried.

The copolymer yields and intrinsic viscosities are also reported in Table I.

TABLE I

Copolymer Yields and Intrinsic Viscosities

RUN NO.	(AA) M	(SUBSTRATE) M	(HNO_3) N	(Ce^{+4}) M	TEMP	REACT. TIME (hrs)	FRACTION YIELD (%)		(η)
21	.0298	0.00241 Amylose	0.066	1.95×10^{-4}	25° C	1	1.0	(21-1)	5.8
							0.77	(21-2)	2.85
25	.0643	0.00241 Amylopectin	0.14	.00338	25° C	30	0.12	(25-1)	--
							11.8	(25-2)	6.7
26	.0645	0.00241 Amylopectin	0.1447	.0002564	25° C	72	65.3	(26-2)	46.5 (5.05)*
							0.3	(26-1)	36.

*In 1.5% NaCl solution

D. Viscosity Measurement

1. Intrinsic Viscosity

[η] was determined by using Ostwald-Fenske capillary viscometers and calculated from extrapolating to zero concentration.

2. Absolute Viscosity Measurement

A Brookfield cone-plate viscometer Model CVT C/P, with an 0.8° cone viscometer, was used.

IV. RESULTS AND DISCUSSION

The first investigation was to determine the cerium (IV) ion/HNO_3 system which would form little or no homopolymer in the presence of the acrylamide monomer, yet would induce a rapid grafting reaction when a suitable substrate was present.

For the experimental section, the various concentration ranges of monomer, nitric acid and cerium (IV) ion used is mentioned. At high acrylamide concentrations (i.e. 6.0 M) homopolymerization was always observed. Also, high nitric acid concentrations resulted in acrylamide homopolymerization. High cerium (IV) ion concentrations (e.g. 0.010 M) in the presence of high (e.g. 6.0 M) acrylamide concentrations also resulted in homopolymer formation. However, acrylamide concentrations of < 2.0 M, with a nitric acid concentration of between 0.056-1.2 M in the presence of 0.01 M cerium (IV) ion resulted in no homopolymerization.

The first substrate investigated was the highly branched amylopectin, a poly(α-D-glucose) with 1-4 linkage and branched at 1-6 position. An aqueous 2% by weight amyleopectin ([η]= 1.18) solution was copolymerized with 0.5% acrylamide four $[Ce^{+4}]$ and three $[H^+]$ concentrations were investigated. Figure 1 shows the final reaction solution viscosities at 24° C after two hours polymerization at 40° C.

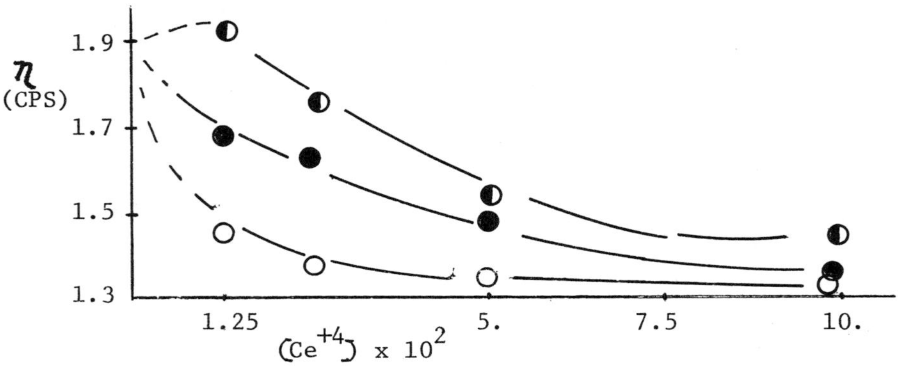

Fig. 1. Concentration of Ce^{+4} effect on amylopectin grafting reaction at 40° C, 0.0655M of $[C_6H_{10}O_5]$, 0.0585M of [AA] (1)"◐" 0.0224 of $[HNO_3]$; (2) "●" 0.048M of $[HNO_3]$; (3) "○" 0.0896 M of $[HNO_3]$.

This figure would seem to suggest only one grafting condition resulting in a greater solution viscosity at that particular monomer to substrate ratio.

These graft copolymers did not precipitate upon cooling as did the unmodified amylopectin solution in the presence of homopolymeric polyacrylamide. Indeed, the graft copolymer was soluble at all temperatures tested (i.e. 0° C to 70° C).

Under certain conditions it has been proposed that the predominant grafting reaction occurs through terminal hemi-acetal cleavage,

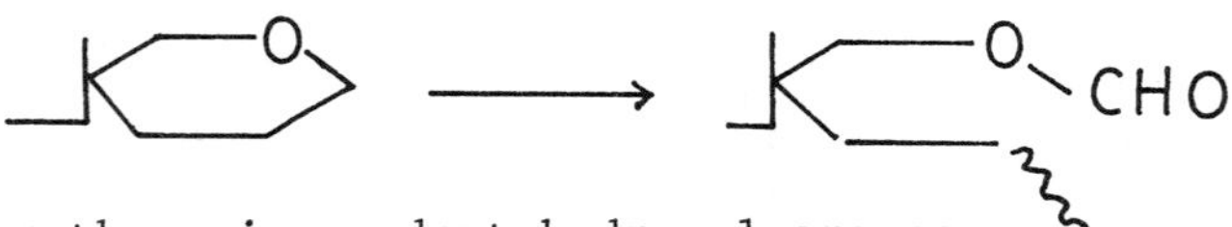

rather than via pendant hydroxyl groups.

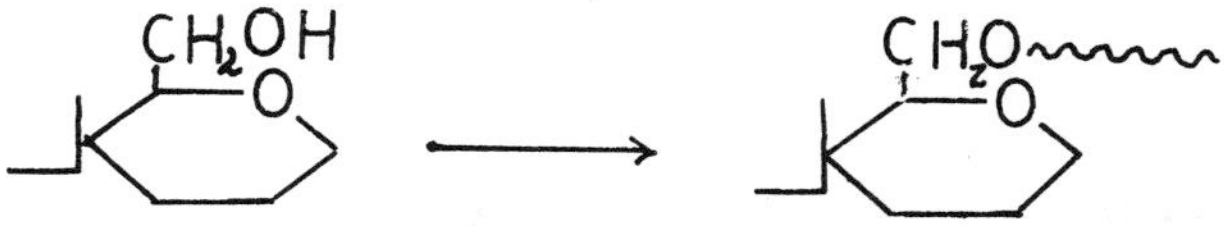

A highly branched species such as amylopectin would be expected to have numerous graft sites, with either mechanism operative. However, linear amylose, which will be discussed later, would be expected to form an AB or ABA block copolymer via terminal hemiacetal cleavage and a graft copolymer, assuming pendant hydroxyl group graft site propagation. Side chain shortly grafting onto an already highly branched substrate would not be expected to significantly alter the mean radius of gyration. Thus, an actual decrease in [η] could be anticipated.

However, growth from the terminal hemiacetal group would be expected to increase the mean radius of gyration, resulting in a corresponding increase in [η]. In the case of the branched amylopectin, because of the numerous terminal groups, very high monomer to substrate ratios would be needed in order to see measurable viscosity increases. However, if any portion of the terminal sites were active graft sites on the amylopectin, smaller monomer to substrate ratios would give measurable viscosity increases. Also, a linear substrate such as amylose would be expected, to be affected by smaller monomer to substrate ratios, if polymerization occurred only via the terminal hemiacetal sites.

As seen in Figure 2, the [η] of the resulting poly(amylopectin-g-acrylamide) demonstrated an apparent maxima. This would be expected, as one could predict from previous statements. Too many active sites can be expected to result in a large number of very short graft segments. These short segments would not be as effective in increasing viscosity as a fewer number of longer segments.

In the case of varying the nitric acid concentration, as seen in Figure 3, one also sees a maxima between 5.5×10^{-2} and 7.5×10^{-2} M for the amylopectin. If the hemiacetal cleavage is the rate determining step, as discussed earlier, it would be anticipated that an optimum pH would exist favoring such a cleavage.

Dextran (m.w. 5-40 million and an [η] of 1.5), a highly branched, mainly poly (1-6-α-D-glucose) was the second substrate investigated. Figures 2 and 3 demonstrate a similar behavior to amylopectin, although a significantly higher $[H^+]$ was needed in order to effect the maximum viscosity.

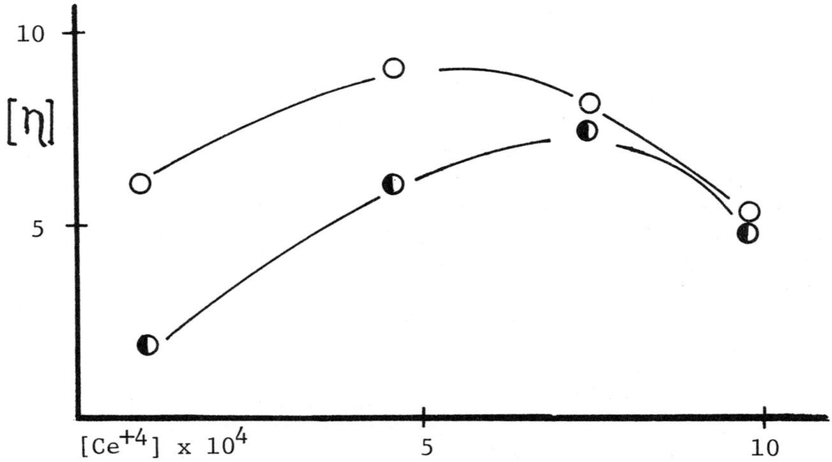

Fig. 2. Ceric ion concentration effect on grafting reaction at 24° C. "O" represents [AA] = 0.992 M, $[HNO_3]$ = 0.05645 M, and $[C_6H_{10}O_5]$ amylopectin = 0.0617 M. "◐" represents [AA] = 0.992 M, $[HNO_3]$ = 0.05645 M, and $[C_6H_{10}O_5]$ dextran = 0.0617 M.

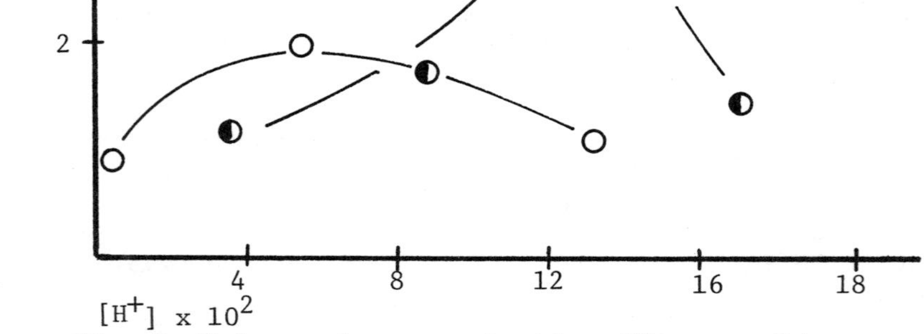

Fig. 3. Hydrogen ion concentration effect, grafting reaction. "◐" represents 0.062 M of $C_6H_{10}O_5$ of dextran, 1.13 M of [AA] and 1.6 x 10^{-4} M of $[Ce^{+4}]$ at 24° C. "O" represents 0.0412 M of $C_6H_{10}O_5$ of amylopectin, 0.376 M of [AA] and 2.7 x 10^{-4} M of $[Ce^{+4}]$ at 24° C.

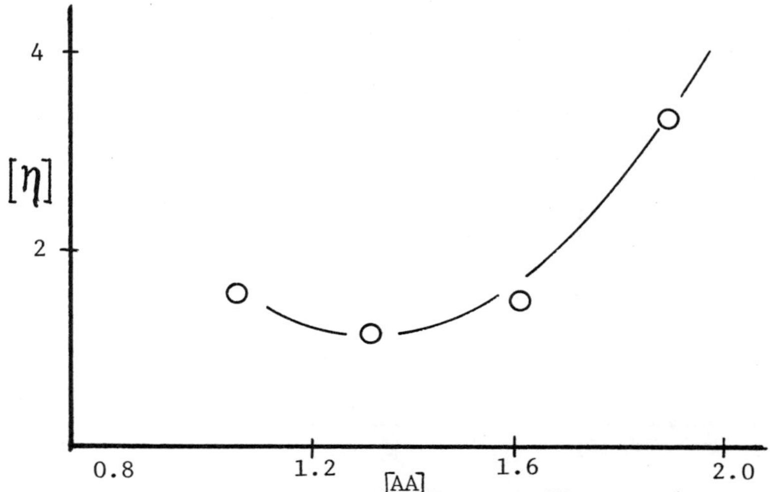

Fig. 4. Acrylamide concentration effect on dextran grafting reaction at 0.062 M of glucose units of dextran, 0.045 M of [HNO_3] and 1.6 x 10^{-4} M of [Ce^{+4}] at 24° C.

The lower [η] at given [Ce^{+4}] can be explained by the more terminal sites expected on the more highly branched dextran, resulting in shorter graft segments.

Figure 4 represents the results of varying the substrate to acrylamide ratio. A plateau is observed where even a four-fold increase in acrylamide concentration results in little or no increase in [η].

One can postulate that at approximately 1.8 M significant graft segment chain lengths are realized. This in turn greatly affects the viscosity.

The linear polysaccharide amylose, poly (1,4-α-D-glucose) was the third substrate investigated. This low molecular weight (e.g. 48,600) polymer had an intrinsic viscosity of 0.2. Figure 5 demonstrates that the grafting of acrylamide onto amylose appears to possess the same sensitivity to cerium (IV) ion and hydrogen ion concentration as did the amylopectin and dextran.

As Table I indicates, a similar viscosity increase was realized at a much lower acrylamide to substrate ratio in the case of the amylose, as compared to the amylopectin.

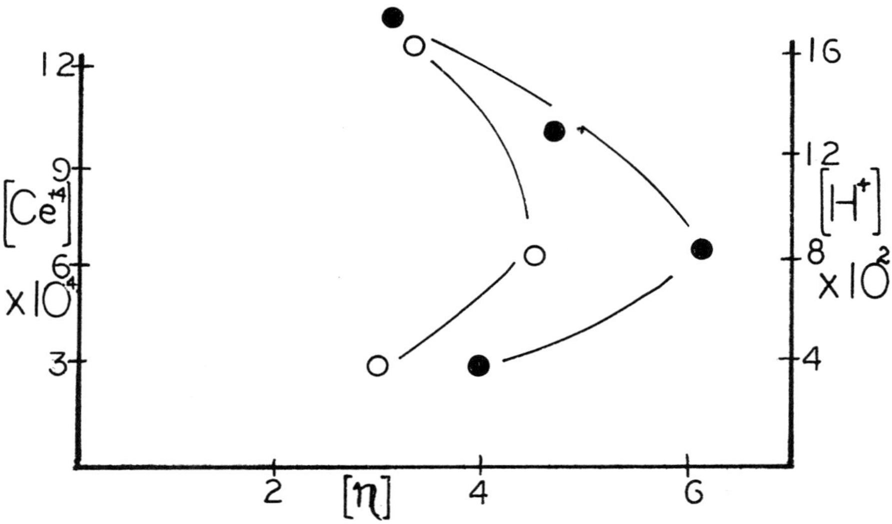

Fig. 5 $[H^+]$ and $[Ce^{+4}]$ effect on amylose grafting reaction. "O" = $[Ce^{+4}]$, "●" = $[H^+]$.

The most recent work on producing grafted amylopectin species has involved the use of sodium acrylate and acrylamide as cografting monomers. A 1 to 7.5 comonomers to amylopectin ratio by weight with the sodium acrylate to acrylamide ratio of 1 to 2, resulted in an [η] of 131 in distilled water and 10.5 in 1.5% NaCl.

V. CONCLUSIONS

Comparing previous studies and the results of this study the order of reactivity of oxidation sites at the end groups of polysaccharides is hemiacetal (or free aldehyde) $>C_2-C_3$, C_3-C_4 or C_4-C_5 diol $>C_6$-OH. Whether or not these grafting sites react appears to depend on the oxidation potential and equilibrium constant and/or disproportionation constant of the oxidant-glucose complexes. The concentrations of ceric (IV) ion and hydrogen ion play very important roles in the reduction-oxidation potential of these systems and affect the grafting reaction, as shown in our results. The acrylamide homopolymerization is also dependent on the concentrations of ceric (IV) ion, hydrogen ion and acrylamide. The equilibrium constant and/or disproportionation constant of the ceric (IV) ion-acrylamide complex may also be important in acrylamide homopolymerization reaction.

References

1. Aggarwal,S.L., "Block Polymers", Plenum press, 1970.
2. Ceresa,R.J., "Block and Graft Copolymers," Butterworths, 1962.
3. Mino,G.,Kaizerman,S.,and Rasmussen,E., J. Amer. Chem. Soc. 81, 1494 (1959).
4. Iwakura,Y.,Kurosaki,T., and Imai,K., J. Polymer Sci., A,3, 1185 (1965).
5. Cumberbirch,R.J.E., and Holker,J.R., J. Soc. Dyers Colourists, 82, 59 (1966).
6. Hintz,H.L., and Johson,D.C., J. Org. Chem., 32, 556 (1967).
7. Duke,F.R., and Forist,A.A., J.Amer. Chem. Soc., 71, 2790 (1949).
8. Terasaki,I. and Matsuki,M. Sen-i-Gakkaishi, 18, 147 (1962).
9. Neimo,L. and Sihtola,H.,Paeri Puu, 6, 369 (1966).
10. Wallace,R.A. and Young,D.G., J. Polymer Sci., A-1, 4,1179 (1966).
11. Duke,F.R. and Bremer,R.F., J. Amer. Chem. Soc., 73, 5179 (1951).
12. Ardon,M., J. Chem. Soc., 1957, 1811.
13. Guibault,G.G. and McCurdy, W.G., J. Phys. Chem., 67, 283 (1963).
14. Muhammad,S.S. and Rao,K.V., Bull. Chem. Soc. Japan, 36, 943 (1963).
15. Venkatakrishnan,S., and Santappa,M., Z. Physik. Chem. (Frankfurt), 16, 73 (1958).
16. Bhargava,K.P., Shanker,R., and Joshi,S.N., J. Sci. Ind. Res.(India), 21B, 573 (1962).
17. Littler, J.S., J. Chem. Soc., 1959, 4135.
18. Ei-Tantawy, Y.A., and Rechnitz,G.A., Anal. Chem., 36, 1774 (1964).
19. Mehrotra,R.N., Z. Physik. Chem.(Leipsig), 230, 221(1965).
20. McAuley,A., J. Chem. Soc., 1965, 4054.
21. Muhammad,S.S., and Rao, K.V., Bull. Chem. Soc., Japan, 36 949 (1963).
22. Bailar,J.C., Ed.,"The Chemistry of Coordination Compounds" p.221, Reinhold, New York, 1956.
23. Cotton,F.A., and Wilkinson,G., "Advanced Inorganic Chemistry," p.547, Interscience, New York, 1962.
24. Chaberek,S., and Martell,A., "Organic Sequestering Agents," p.140, Wiley, N.Y., 1959.
25. Hintz,H.L., Dissertation, The Institute of Paper Chemistry, Appleton, Wisc., 1966.
26. Mehrotra,R.N., and Amis,E.S., J. Org. Chem., 39, 1788 (1974).

27. Pottenger,C.R., and Johnson,D.C., J. Polymer Sci.,A-1,8, 301 (1970).
28. Cantor,S.M., and Peniston,O.P., J.Amer. Chem. Soc., 62, 2113 (1940).
29. Mino,G., and Kaizerman,S., J. Polymer Sci., 31, 242(1958).
30. Mino,G., Kaizerman,S., and Rasmussen,E., J. Polymer Sci., 38, 393 (1959).
31. Arthur,J.C.,Baugh,P.J., and Hinojosa, O., J. Appl. Polymer Sci., 10, 1591(1966).
32. Takahashi,T., Hori,Y., and Sato,I., J. Polymer Sci., A-1, 6, 2091 (1968).

THE EFFECT OF SOLVENTS ON POLYMERIZATION RATE, PARTICLE SIZE, AND MOLECULAR WEIGHT DISTRIBUTION IN THE HETEROGENEOUS SOLUTION POLYMERIZATION OF VINYL ACETATE

G. ALLAN STAHL, Department of Chemistry, University of Alabama, and RAYMOND B. SEYMOUR, Department of Polymer Science, University of Southern Mississippi.

I. INTRODUCTION

Heterogeneous polymerization will occur if the difference between the solubility parameter of a polymer and the solvent in which it is prepared is greater than 1.8 Hildebrand units (H) (1). It has also been shown that the rates of initiation and propagation in these heterogeneous solvent systems are essentially unaltered, but the rate of termination is drastically decreased (2). Thus, occluded macroradicals with characteristic long lives are produced. Presumably, the polymer molecule coils and protects the growing macroradical from termination. Accordingly, block copolymers have been produced by adding additional vinyl monomers to these macroradicals (1,3,4,5,6).

Long lived macroradicals have also been produced in viscous media by the well known gel effect. It has been shown that the rates of initiation and propagation are unchanged, but the rate of termination is decreased (7).

Slavnitskaya, et al., (8) polymerized methyl methacrylate in cyclohexane, and used a scanning electron microscope to measure the size of the particles formed in the reaction. Particles about 1 μ in size were formed at higher monomer concentrations, while in the less solvating medium of lower monomer concentration, the size of the particles decreased to 0.1 μ.

In this study, poly(vinyl acetate) (PVAc) was prepared by heterogeneous solution polymerization. The effect of changing the solvency and viscosity of the medium on the polymerization rate, particle size, and molecular weight distribution has been studied. The polymerization rates were monitored by precision dilatometry. The particle size and number was determined by scanning electron microscopy, and the molecular weight distributions were determined by gel permeation chromatography.

II. EXPERIMENTAL

Vinyl acetate (VAc) was purified by double distillation at reduced pressure after drying over anhydrous Na_2SO_4 for 48 hours. Poly(dimethylsiloxane) or silicone oil (Dow-Corning 200 Fluid), and perfluorohexane, perfluorooctane, and perfluoro-2-butyl-tetrahydrofuran (3-M Corporation) were used as received. All other solvents were reagent grade (J. T. Baker and Co.) and used without further purification.

Relative polymerization rates were determined by use of a previously described recording capacitance dilatometer (9). In the dilatometer 4 ml of vinyl acetate (VAc) with two percent tert.-butyl peroxypivalate (Lupersol 11[1], Lucidol Div., Pennwalt Corp., 75% solution in mineral spirits, half-life 12h at 50°C) was heated at 50°C in 50 ml solvent for 24h.

PVAc was prepared for scanning electron microscopy studies by heating, at 50°C, 1.0 ml of VAc and 20 ml of solvent with 2.30, 3.45, 4.60, or 5.75 x 10^{-3}M tert.-butyl peroxypivalate in 1 oz. brown bottles with foil-lined phenolic screw caps. A 1 x 4 cm aluminum foil strip was included in each bottle. After 24h, the strip was removed from the solvent. The polymer which precipitated on the strip was dried, and then coated with Au/Po with a Technics Hummer.

A Cambridge Stereoscan Model S410 scanning electron microscope was employed to determine particle size and number. The diameter of the microparticles was measured directly on the microscope monitor screen. The size and number of PVAc particles produced in silicone oil was determined by first diluting to 100 ml with hexane. A film of 0.01 ml of this solution was cast, and the number of particles estimated by viewing a random area of the film.

Gel-permeation chromatograms were obtained by using a GPC/ALC Waters Model 301 gel permeation chromatograph. Molecular weights were determined by comparison with calibrated Waters Associates standard polystyrene samples. Viscometric data were obtained by averaging five effluent times measured to ±0.1 seconds at 30.0°C in a No. 1 Ubbelohde viscometer.

Diffracted scanning calorimetry was performed on a DuPont Model 990 DSC thermal analyzer.

III. DISCUSSION

PVAc has been shown to chain transfer readily with many common organic solvents. For this reason cyclohexane (10) and tert.-butanol (11) were chosen as poor and good solvents, respectively. As shown in Figure 1, the rate of VAc

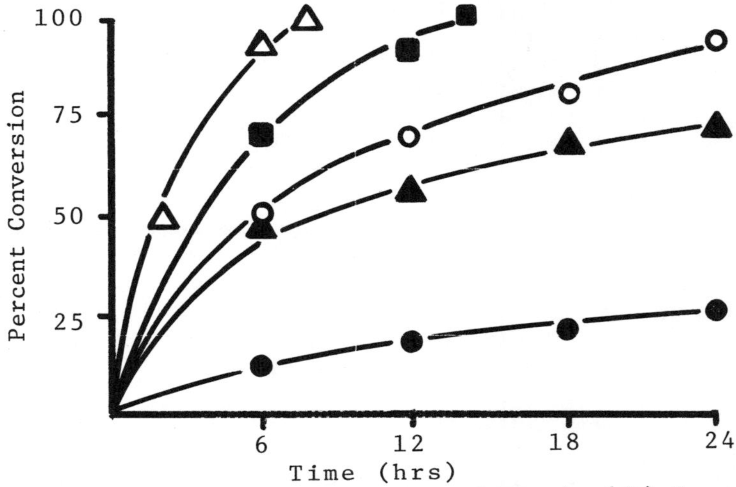

Figure 1. The Rate of Polymerization of VAc in (●) Benzene, (▲) tert.-Butanol, (○) Cyclohexane, (■) Cyclohexane plus 5% fumed silica, (△) tert.- Butanol plus 5% fumed silica.

polymerization is faster in tert.-butanol than in benzene, which is also a good solvent for PVAc. The rate of polymerization is even faster in cyclohexane. In the poor solvent, the polymer coils and occludes the growing macroradical. Since the gel effect also reduces the rate of termination of macroradicals, the rate of polymerization of VAc was increased when fumed silica was added to tert.-butanol. An increase in the rate of polymerization of VAc was also noted when fumed silica was added to cyclohexane.

The increase in rate of polymerization in viscous poor solvents is shown in Figure 2. Dow-Corning 200 Fluid (silicone oil) was selected as the viscous poor solvent because it has a low rate of chain transfer with VAc macroradicals, and is commercially available in a wide range of viscosities. The rate of formation of PVAc is rapid in 50cs silicone oil, yet is faster in 200cs silicone oil, and the fastest in the more viscous system, 500cs silicone oil.

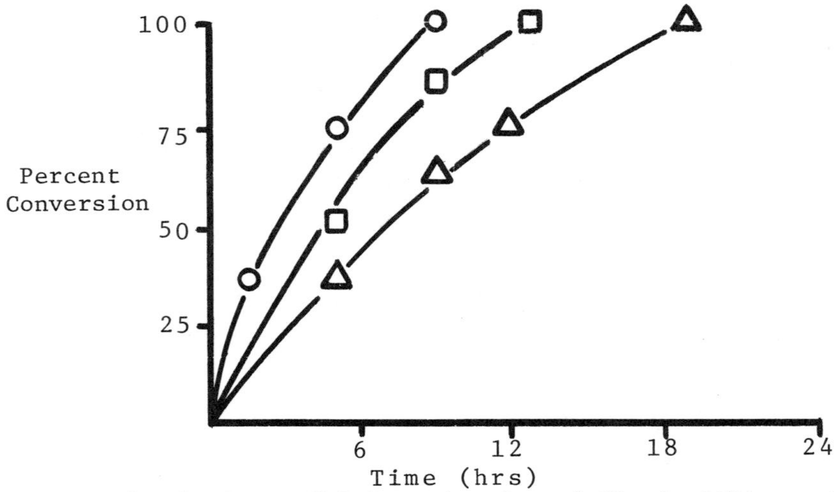

Figure 2. The Rate of Polymerization of VAc in Silicone Oil of Different Viscosities at 50°C. (Δ) 50cs, (□) 200cs, and (O) 500cs.

As shown in Figure 3, uniform particles of PVAc with a diameter of 230 ± 10 nm were obtained in silicone oil regardless of viscosity. The difference in solubility parameter between PVAc (δ = 9.4 H) and silicone oil (δ = 6.6 H) is 2.8 H.

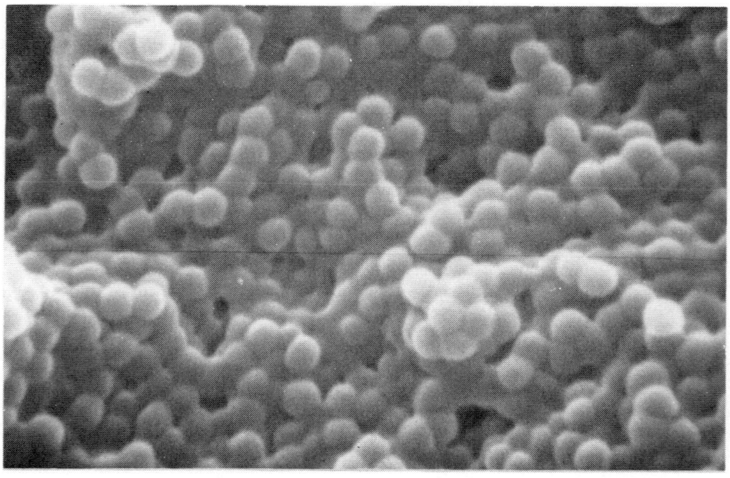

Figure 3. Scanning Electron Micrograph of PVAc Produced in 500 cs Silicone Oil at 50°C. (Magnification = 21,000 X).

The dependence of particle size on the difference in solubility parameter is shown in Table I.

Table I
Relationship of the Size of Poly(Vinyl Acetate) Particles and the Difference in Solubility Parameter Between Polymer and Solvent.

Solvent	$\delta_{sol.}H$	$\lvert\delta_p - \delta_s\rvert$	Particle size
perfluorohexane	5.6	3.8	170 ± 10 nm
perfluorooctane	5.7	3.7	170 ± 10 nm
perfluoro-2-butyl-tetrahydrofuran	6.0	3.4	180 ± 10 nm
silicone oil	6.6	2.8	230 ± 10 nm
hexane	7.3	2.1	4 - 10 μ
heptane	7.4	2.0	10 - 20 μ
methylcyclohexane	7.8	1.6	10 - 30 μ[a]
cyclohexane	8.2	1.2	50 - 800 μ[a]

a. polymer particles showed extensive agglomeration

PVAc particles prepared in solvents with large differences in solubility parameter compared to the macro-radical are smaller. The particles are a uniform 170 nm in perfluorohexane and perfluorooctane. In perfluoro-2-butyl-tetrahydrofuran and silicone oil, the particles are slightly larger, 180 and 230 nm, respectively. As the difference in solubility parameter approaches polymer solvation, the range of particle sizes as well as the diameter become greater. Particles in hexane and haptane are 4-10 μ and 10-20 μ, respectively. Particles produced in methyl cyclohexane and cyclohexane have agglomerated, and are in the range of 10-30μ and 50-800 μ.

Particle formation starts with the production of radicals by the decomposition of an initiator, which in turn reacts with the monomer to form a growing oligomeric chain. Fitch and Tsai (12) proposed that each oligomer chain at first has an extended configuration in solution, but then collapses into a condensed state when it reaches a certain threshold molecular weight, depending on its solubility in the medium. This condensed oligomeric chain constitutes a new particle nucleus. The behavior of each oligomer chain is unaffected by the presence of other oligomeric molecules. Therefore each oligomer chain forms a new particle unless it is captured by an existing particle before it reaches the threshold molecular weight. Fitch and Tsai predict the formation of all the particles within a short time of initiation.

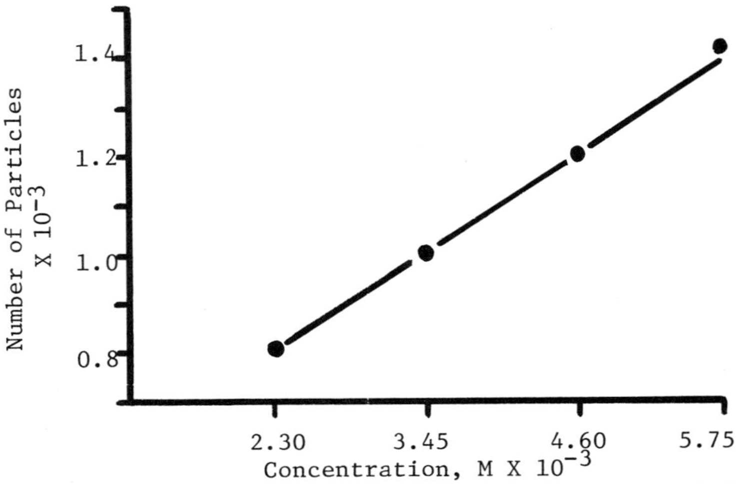

Figure 4. The Relationship of the Number of PVAc Particles Prepared in 500cs Silicone Oil at 50°C and the Initial Concentration of Tert.-Butyl Peroxypivalate.

In Figure 4, the number of particles produced when vinyl acetate is polymerized in silicone oil (500 cs) is compared with the initial molar concentration of the initiator (tert. butyl peroxypivalate). A linear relationship can be drawn. This infers that the Fitch and Tsai self nucleation is operative. However, as shown in Figure 5, the particles produced in silicone oil (500 cs), when the initiator concentration is varied, remain a uniform 230 ± 10 nm.

As shown by the gel permeation chromatograms in Figure 6, the molecular weight is decreased and becomes more dispersed as the difference in solubility parameters ($|\delta_p - \delta_s|$) becomes less. Curve A is high molecular weight (Mw = 121,500) and narrow dispersed (Mw/Mn = 2.48) PVAc produced in silicone oil. PVAc produced in hexane, curve B, is lower molecular weight (Mw = 82,000) and broader dispersed (Mw/Mn = 3.28). Curve C shows PVAc produced in heptane. The molecular weight is lower (Mw = 64,500) and more polydispersed (Mw/Mn = 4.03). Table II lists the molecular weights and dispersities of PVAc produced in various solvents.

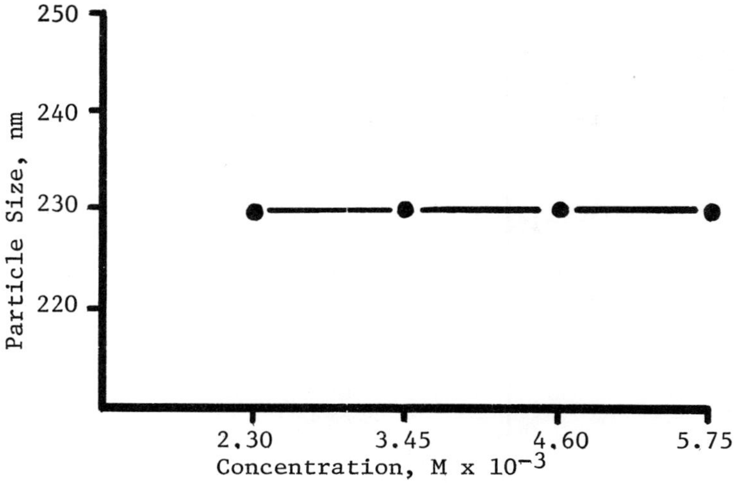

Figure 5. The Relationship of the Size of PVAc Particles Prepared in 500 cs Silicone Oil at 50°C and the Initial Concentration of Tert.-Butyl Peroxypivalate.

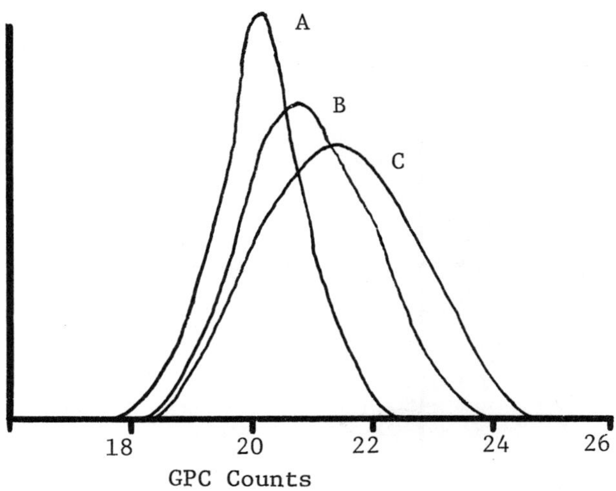

Figure 6. Gel Permeation Chromatogram of PVAc Produced in (A) Silicone Oil, (B) Hexane, and (C) Heptane at 50°C.

Table II
Molecular Weight Data for PVAc Produced in Various Solvents at 50°C

Solvent	$\lvert\delta_p - \delta_s\rvert$	Mw	Mn	Mw/Mn
perfluorohexane	3.8	69,500	24,500	2.84
perfluorooctane	3.7	71,000	25,500	2.78
perfluoro-2-butyl-tetrahydrofuran	3.4	78,500	28,000	2.80
silicone oil	2.8	121,500	49,000	2.48
hexane	2.1	82,000	25,000	3.28
heptane	2.0	64,500	16,000	4.03
methyl cyclohexane	1.6	55,000	12,500	4.40
cyclohexane	1.2	83,000	21,000	3.96

The purity of PVAc was confirmed by infrared spectroscopy and differential scanning calorimetry (DSC). A typical DSC thermogram is shown in Figure 7.

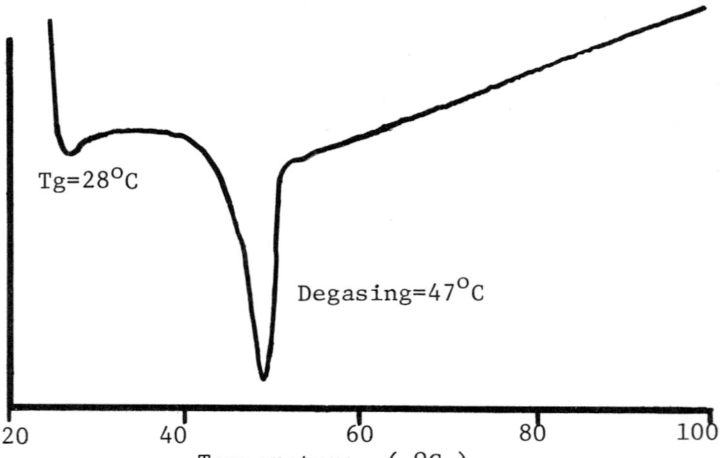

Figure 7. Differential Scanning Calorimeter Thermogram of PVAc Prepared in 500 cs Silicone Oil at 50°C.

IV. CONCLUSIONS

The rate of polymerization and size of the particles produced, when VAC is polymerized in poor solvents, is

proportional to the difference in solubility parameter ($\Delta\delta$) between the polymer and solvent. The number of particles was found to be proportional to the initiator concentration.

The rate of polymerization is also proportional to the solution viscosity, but the particle size is independent of solution viscosity and initiator concentration. The rapid rates of polymerization noted in a viscous poor solvent can be explained if one considers a particle with occluded macroradicals, suspended or dispersed by the viscosity of the medium, thereby having more surface area exposed to reactant rich diluent. In nonviscous poor solvents agglommeration can occur burying radicals deep in solids and increasing the time required for diffusion of reactants.

The authors wish to thank Mr. Robert Kieth, Department of Electrical Engineering, University of Houston, Houston, Tx, for the scanning electron micrographs. This work was supported, in part, by the ACS/Petroleum Research Fund.

V. BIBLIOGRAPHY

1. Seymour, R. B., Owen, D. R., and Stahl, G. A., Polymer, 14, 324 (1973).

2. Bamford, C. H. and Jenkins, A. D., Proc. Roy. Soc. (London), A216, 515 (1953).

3. Seymour, R. B., Owen, D. R., Stahl, G. A., Wood, H. and Tinnerman, W. N., Applied Polymer Symp., 25, 69 (1974).

4. Seymour, R. B., Stahl, G. A., Owen, D. R., and Wood, H., Advan. Chem. Ser., 142, 309 (1975).

5. Seymour, R. B., Stahl, G. A., and Wood, H., Applied Polymer Symp., 26, 249 (1975).

6. Seymour, R. B. and Stahl, G. A., J. Polymer Sci., Chem. Ed., 14, 000 (1976).

7. Trommsdorff, E., Kohle, H., and Legally, P., Makromol. Chem., 51, 154 (1962).

8. Slavinitskaya, N. N., Semchikov, Y. D., Ryabov, A. V., and Bort, D. N., Vysokomol. Sved., A1Z, 1756 (1970); translated in Polym. Sci. USSR, 12, 1993 (1970).

9. Seymour, R. B., and Stahl, G. A., Rev. Sci. Instrum., 46, 1467 (1975).

10. Ham, G. E., "Vinyl Polymerization, Part I," Marcel Dekker, New York, 1967, pp 208-329.

11. Minsk, L. M. and Taylor, E. W., U. S. Pat. 2,582,055 (January 8, 1952).

12. Fitch, R. M. and Tsai, C. H., in "Polymer Colloids," R. M. Fitch, Ed., Plenum Press, New York, 1971. pp 73-103.

INDEX

A
B 7
C 8
D 9
E 0
F 1
G 2
H 3
I 4
J 5

A
B 7
C 8
D 9
E 0
F 1
G 2
H 3
I 4
J 5